First Edition

Fire Department Safety Officer

Fred Stowell, Writer

Cynthia Brakhage, Editor

Carol Smith, Editor

Validated by the
International Fire Service Training
Association

Published by
Fire Protection Publications,
Oklahoma State University

Photo courtesy of Chris Mickal, New Orleans Fire Department.

The International Fire Service Training Association

The International Fire Service Training Association (IFSTA) was established in 1934 as a "nonprofit educational association of fire fighting personnel who are dedicated to upgrading fire fighting techniques and safety through training." To carry out the mission of IFSTA, Fire Protection Publications was established as an entity of Oklahoma State University. Fire Protection Publications' primary function is to publish and disseminate training texts as proposed and validated by IFSTA. As a secondary function, Fire Protection Publications researches, acquires, produces, and markets high-quality learning and teaching aids as consistent with IFSTA's mission.

The IFSTA Validation Conference is held the second full week in July. Committees of technical experts meet and work at the conference addressing the current standards of the National Fire Protection Association and other standard-making groups as applicable. The Validation Conference brings together individuals from several related and allied fields, such as:

- Key fire department executives and training officers
- Educators from colleges and universities
- Representatives from governmental agencies
- Delegates of firefighter associations and industrial organizations

Committee members are not paid nor are they reimbursed for their expenses by IFSTA or Fire Protection Publications. They participate because of commitment to the fire service and its future through training. Being on a committee is prestigious in the fire service community, and committee members are acknowledged leaders in their fields. This unique feature provides a close relationship between the International Fire Service Training Association and fire protection agencies which helps to correlate the efforts of all concerned.

IFSTA manuals are now the official teaching texts of most of the states and provinces of North America. Additionally, numerous U.S. and Canadian government agencies as well as other English-speaking countries have officially accepted the IFSTA manuals.

ISBN 0-87939-191-x   *Library of Congress LC# 00-112095*

First Edition, First Printing, February 2001   *Printed in the United States of America*
10   9   8   7   6   5

If you need additional information concerning the International Fire Service Training Association (IFSTA) or Fire Protection Publications, contact:

Customer Service, Fire Protection Publications, Oklahoma State University
930 North Willis, Stillwater, OK 74078-8045
800-654-4055   Fax: 405-744-8204

For assistance with training materials, to recommend material for inclusion in an IFSTA manual, or to ask questions or comment on manual content, contact:

Editorial Department, Fire Protection Publications, Oklahoma State University
930 North Willis, Stillwater, OK 74078-8045
405-744-4111   Fax: 405-744-4112   E-mail: editors@osufpp.org

Table of Contents

Preface

This first edition of the **Fire Department Safety Officer** manual is intended to replace the **Fire Department Occupational Safety** manual, second edition. The information provided in IFSTA's new **Fire Department Safety Officer** manual addresses NFPA 1521, *Standard for Fire Department Safety Officer*, which identifies the performance requirements necessary to fulfill the duties of a fire department safety officer. This manual also covers the applicable performance requirements of NFPA 1500, *Standard on Fire Department Occupational Safety and Health Program*, which contains the minimum requirements and procedures for a safety and health program, and NFPA 1021, *Standard for Fire Officer Professional Qualifications*, which establishes the minimum requirements for a fire department officer.

Firefighter health and safety should be a basic consideration in all fire service operations and activities. Increased efforts in the area of health and safety can reduce line-of-duty death and injury, reduce long-term disability, and improve firefighter morale and quality of life. The health and safety officer and the incident safety officer are essential to the goals of improved firefighter health and safety. This manual is designed to provide the health and safety officer and the incident safety officer with the information necessary to perform the various tasks assigned to them and to provide direction to information and resources that will further their knowledge of safety-related issues.

Acknowledgment and special thanks are extended to the members of the validation committee who contributed their time, wisdom, knowledge, and guidance to the creation of this manual.

Committee Chair
Jeffrey Morrissette
State Fire Administrator
Commission on Fire Prevention and Control
Windsor Locks, CT

Secretary
Fire Chief Stephen Ashbrock
Madiera and Indian Hill Fire Department
Cincinnati, OH

Firefighter Bill Bingham
V.A. Medical Center
Bell Buckle, TN

Firefighter/EMT Paul H. Boecker III
Sugar Grove Fire District
Oswego, IL

Captain Keith Boswell
Justice Institute Fire & Safety Division
New Westminster, BC, Canada

Fire Chief Bradd K. Clark
Sand Springs (OK) Fire Department
Sand Springs, OK

Donald G. Elrod
Manager, Emergency Management
USEC
Paducah, KY

Stephen Foley
Sr. Fire Service Specialist
National Fire Protection Association
Quincy, MA

Assistant Chief Robert Giorgio
Cherry Hill (NJ) Fire Department
Cherry Hill, NJ

Charles Holman
Fire Training Manager
Palisades Nuclear Plant
Covert, MI

Fire Chief Michael R. Linville
Sunoco Refinery Fire Department
Tulsa, OK

Captain Murrey Loflin
Virginia Beach (VA) Fire Department
Virginia Beach, VA

Fire Chief Rick D. McIntyre
Jacksonville (NC) Fire Department
Jacksonville, NC

Firefighter Jarett Metheny
Midwest City (OK) Fire Department
Midwest City, OK

Richard Pippenger
Resource Officer
Lower Valley Fire Dist.
Grand Junction, CO

Assistant Chief Cliff Puckett
Mesa (AZ) Fire Department
Mesa, AZ

Fire Chief J. D. Rice
Valdosta (GA) Fire Department
Valdosta, GA

David Ross
Chief Health and Safety Officer
Toronto (ON) Fire Services
Toronto, Ontario, Canada

Charles C. Soros
Western Regional Director
Fire Dept. Safety Officers Association
Seattle, WA

Fire Marshal Paul Valentine
Mt. Prospect (IL) Fire Department
Mt. Prospect, IL

Contract Writer
Fred Stowell
Tulsa, OK

Special appreciation is also extended to the following fire departments that provided their personnel, resources, and time to assist our staff in shooting many of the new photographs that were necessary to complete this manual.

BRADLEY INTERNATIONAL AIRPORT (CT) FIRE DEPARTMENT
Fire Chief John Duffy
Captain Guy Henry
FF Christy Delvey
FF Billy Cockfield
FF Scott Ardel

CONNECTICUT FIRE ACADEMY
Ed O'Hurley
Mike Jepeal
John Kowalski
William Cesareo
John J. Gamble
FF Joe Steward
FF Mark Masterjoseph
FF Mike O'Leary
FF Chad Armstrong
FF Joe Maida

FF Kevin Goodwin
FF Josh Recker
FF John Burelle
FF Ray Sanchez
FF Richard Clow
FF Aaron Lapointe
FF Clint Marth
FF Brian Jennes
FF Matthew Hannon
FF Chris Lioti
FF Leo Pon

HARTFORD (CT) FIRE DEPARTMENT
Fire Chief Charles Teale
Machine Shop Superintendent Jack Regier
Equipment Maintenance Technician Sal Pagliarello
Lieutenants Peter Murphy & David Serpliss
Ladder Driver Thomas Cosgrove
Pump Operator Hilario Sanchez
FF Hermine Hastings

HESC-HEALTHCOMP EVALUATION SERVICES CORPORATION
Marvin Ward, M.A.
Mark Bean, M.A.
Nancy Barrios, M.A.
Kevin Ryan, O.D.
Mary Troiano, L.P.N.
George Pusko, P.A.
Edward Smith
Jamie Thom

TOWN OF MANCHESTER (CT) FIRE RESCUE/EMS
Fire Chief Thomas Weber
D/C Mark Salafia
Captain Heather Burford
Secretary Barbara Mozzer
FF Christian P. Frezza
FF Dave Casellini
FF Jay Gonzalez
FF John McGree
FF Brian Hurst

TULSA (OK) FIRE DEPARTMENT
Mike Mallory, Safety Officer

UNIVERSITY OF CONNECTICUT HEALTH CENTER FIRE DEPARTMENT
D/C Robert Howe
Lieutenant Phil Roche

FF Chris Brewer

WETHERSFIELD (CT) FIRE DEPARTMENT
Chief William Clark
Captain Chuck Flynn
Lt. Brian Schroll
Lt. Mark Guerrera
FF Anthony Dignoti
FF Scott Roberts

The following individuals and organizations contributed information or photographs or otherwise provided assistance that made the completion of this manual possible:

Angel-Guard Products, Inc.
Bob Esposito, Trucksville, Pennsylvania
Fyrepel, Inc.
Martin Grube, Virginia Beach Fire Department
Ron Jeffers, Union City, New Jersey
Mike Mallory, Safety Officer, Tulsa (OK) Fire Department
Captain Chris Mickal, New Orleans (LA) Fire Department Photo Unit
Monterey County, California Fire Training Officer's Association
Jeffrey Morrissette, State Fire Administrator, Commission on Fire Prevention and Control, Windsor
    Locks, CT
Mike Nixon, Portland (ME) Fire Department
SKEDCO, Inc.
Russell Strickland, Maryland Fire & Rescue Institute, College Park, MD
Underwriters Laboratories, Inc.
Weis American Fire Equipment Company
Joel Woods, University of Maryland Fire & Rescue Institute, College Park, MD

Gratitude is also extended to the following members of the Fire Protection Publications staff, whose contributions made the final publication of the manual possible:

Mike Wieder, Manager of IFSTA projects
Barbara Adams, Associate Editor
Don Davis, Coordinator, Publications Production
Ben Brock, Senior Graphic Designer
Ann Moffat, Graphic Design Analyst
Desa Porter, Senior Graphic Designer
Tara Gladden, Editorial Assistant
Lee Noll, Research Technician

Introduction

During the last 25 years of the 20th century, health and safety for firefighters has become increasingly important. Firefighter line-of-duty deaths have averaged close to 100 members per year, and an average of 55,000 annual injuries have occurred at incidents. To combat this staggering toll, the National Fire Protection Association (NFPA) published NFPA 1500, *Standard on Fire Department Occupational Safety and Health Program* (1997 edition), which establishes the minimum requirements for a fire-service-related occupational safety and health program. Within this standard are the requirements for the position of health and safety officer who was given responsibility for managing the program. NFPA 1501, *Standard for Fire Department Safety Officer*, was revised in 1992 and renumbered as NFPA 1521.

As a result of these changes, fire departments in North America have placed an increasing emphasis on firefighter health, safety, and wellness. The health and safety officer's position has become an essential part in the team approach to developing a firefighter health and safety program. Working with members of the department administration, the fire department physician, and representatives of the member organization, the health and safety officer must ensure that safety issues are addressed in the department's health and safety program. When developing the health and safety program, the health and safety officer must be concerned with the following items:

- Techniques of risk management
- Knowledge of applicable safety laws, codes, and standards
- Health and safety training issues
- Incident or accident prevention
- Incident or accident investigation and analysis
- Records management and data analysis

- Apparatus design, operation, care, and maintenance
- Equipment design, operation, care, and maintenance
- Facility design, care, and maintenance
- Infection control
- Critical incident stress management
- Personal protective clothing and equipment design, use, care, and maintenance
- Firefighter health, wellness, and physical fitness
- Emergency incident operations

In addition to the responsibilities of the health and safety officer, NFPA 1521 outlines the duties of the incident safety officer. This position is part of the incident management system and an important role in all emergency incidents. Duties of the incident safety officer may be assigned to a qualified line officer or to the health and safety officer. Some of the responsibilities of this position are as follows:

- Incident scene safety
- Emergency incident accountability system
- Deployment of the rapid intervention crew or team
- Infection control at emergency medical operations
- Safety at fire suppression operations
- Safety at hazardous materials operations
- Safety at special operations
- Postincident analysis
- Incident or accident investigation and analysis

The health and safety program within an emergency response organization requires the following components:

- Commitment by management and members

- Written health and safety program

- Procedures for ongoing review of the program

- Health and safety officer who is trained in the duties of the position and given the authority to perform those duties

Fire departments that meet these requirements provide their members with a safer working environment, a longer, healthier life, and less lost-time injuries and workers compensation claims.

 ## Purpose and Scope

This first edition of **Fire Department Safety Officer** replaces the second edition of **Fire Department Occupational Safety**. This manual is intended to assist the company officer who is assigned the duties and responsibilities of a fire department health and safety officer and incident safety officer. It helps the health and safety officer become more aware of safety issues and regulatory guidelines in the everyday management of the fire department's safety and health program. Some of these duties include preventing or reducing incidents involving fatalities, injuries, exposures and enhancing the overall wellness of firefighters. **Safety Officer** provides direction to the incident safety officer functioning within the incident management system, identifying and preventing unsafe acts, and correcting unsafe conditions at an emergency scene.

This manual addresses the 1997 edition of NFPA 1521, *Standard for Fire Department Safety Officer*. NFPA 1521 contains the suggested qualifications, roles, and responsibilities of the safety officer. The manual also addresses NFPA 1500, *Standard on Fire Department Occupational Safety and Health Program*, as it pertains to the duties and responsibilities of the fire department safety officer in establishing a safety and health program and in performing duties as an incident safety officer.

NOTE: At the time this manual was produced, NFPA was considering eliminating the NFPA 1521 standard. The plan included dividing the information in NFPA 1521 between NFPA 1500 and NFPA 1021, *Standard for Fire Officer Professional Qualifications*. The final decision on this proposal was not available at the press time of this manual.

 ## Notice on Gender Usage

In order to keep sentences uncluttered and easy to read, this text has been written using the masculine gender, rather than both the masculine and female gender pronouns. Years ago, it was traditional to use masculine pronouns to refer to both sexes in a neutral way. This usage is applied to this manual for the purposes of brevity and is not intended to address only one gender. Please note that the included photos and artwork reflect the diversity of today's firefighters.

1

The Evolution of the Fire Department Safety Officer

Chapter 1

The Evolution of the Fire Department Safety Officer

The fire service in North America has traditionally had the reputation of a heroic profession. Images of firefighters risking their lives to save life and property can be found in 18th, 19th, and early 20th century lithographic prints and contemporary news media photographs (Figure 1.1). However, this reputation has come at a cost. For many years, fire fighting had the highest casualty rate of any profession in the country. As late as 1999, line-of-duty activities accounted for 112 firefighter deaths while deaths between 1977 and 1996 averaged 119 per year. The traditional reputation appears justified when you add the reduced life expectancy of firefighters due to exposure to smoke and other hazardous aspirants as well as heart disease due to stress.

In reality, death and injury are not a necessary part of fire fighting. As North America became more industrialized in the last half of the 19th century and technology improved at an increasingly fast pace,

Figure 1.1 Hazards facing firefighters have increased during the past century. *Courtesy of Bob Esposito.*

workplace safety and health became critical issues. Many employers focused strictly on productivity and profit with little or no regard to workers' safety and health. If an employee was injured or disabled, the employee was replaced, receiving little or no compensation. Even less was provided to relatives if an employee suffered a fatal occupational accident. Most employers were exonerated of any criminal or civil penalties because few laws relating to compensation or employer liability existed at the time.

The fire service was ready to acknowledge the level of risk by the mid-twentieth century. To allow firefighters more time to recuperate, some career fire departments shifted from a one-day-on, one-day-off work schedule to a one-day-on, two-day off-schedule. City and state governments developed better benefit and retirement packages for firefighters. These pensions and benefits, along with the heroic image, made the fire service an attractive career to many men and women. At the same time, increased hazards created by new manufacturing technology were becoming more prevalent. Exposure to lethal by-products of combustion and hazardous chemicals increased the risk to firefighters. Increased responsibility for medical emergencies exposed more firefighters to communicable diseases. Although some departments worked to improve respiratory protection, hearing protection, and protective clothing beginning in the 1950s and 1960s, the long-term effects of stress, heart and lung disease, and hearing loss on personnel were not fully appreciated. Change in the work schedule and shortened service times did not mitigate all the inherent safety hazards. In general, the fire service did not take a proactive stance to protect its members (Table 1.1).

This chapter describes the regulatory requirements for the fire department's health and safety program, the appropriate NFPA standards, the need for health and safety and incident safety officers, the qualifications for personnel holding these positions, and the authority necessary for these officers to perform their duties.

 ## Occupational Safety and Health Act

In December, 1970, President Richard Nixon signed into law the Occupational Safety and Health Act of 1970 (Public Law 91-596). This law is commonly referred to as the Williams-Steiger Occupational Safety and Health Act, named after the two legislators who were the primary authors and sponsors. The Act instructed the Secretary of the U.S. Department of Labor to promulgate mandatory occupational safety and health standards, among other requirements. The creation of the Occupational Safety and Health Administration (OSHA)

Table 1.1
Statistics on Firefighter Line-of-Duty Fatalities and Injuries

Year	Line-of-Duty Deaths	Fireground Injuries	Non-fireground-Injuries
1987	131	57,755	13,940
1988	136	61,790	12,325
1989	118	58,250	12,580
1990	107	57,100	14,200
1991	108	55,830	15,065
1992	75	52,290	18,140
1993	78	52,885	16,675
1994	104	52,875	11,810
1995	97	50,640	13,500
1996	96	45,725	12,630
1997	97	40,920	14,880
1998	91	43,080	13,960
1999	112	45,500	13,565

These figures are based on data published in the *NFPA Journal* for the period 1987-1999.

under the existing U.S. Department of Labor was a result of this Act. (Canadian readers should refer to Appendix A, Health and Safety Regulations in Canada, for an overview of Canadian occupational health and safety regulations.)

Prior to the initiation of OSHA, there were few national safety and health requirements for employee safety in North America. Numerous states had safety and health regulations that varied in quality and enforcement from excellent to poor. Under the OSHA regulations, the employers covered by the law were required to do the following:

• Furnish each employee a place of employment free from recognized hazards that would cause or likely cause death or serious physical harm.

• Comply with occupational safety and health standards promulgated under PL 91-596.

In addition, Section 5 (b) of the Act required all employees to comply with the "standards, rules, regulations, and orders issued pursuant to this Act."

Although primarily concerned with safety in the private sector, sections of this law were found applicable to the public sector as a whole. The fire service was required to comply with all applicable OSHA regulations published in Title 29 of the *Code of Federal Regulations* (*CFR*) or equivalent regulations issued by state governments that elected to assume responsibility for the development of standards and enforcement as allowed by Section 18, State Jurisdiction and State Plans, of the OSHA Act. Although Title 29 *CFR* 1910, Subpart L, Fire Protection Industrial Fire Brigades, did not apply directly to municipal fire departments or fire protection districts, it contained applicable requirements for the fire service. This standard or regulation contained requirements for personal protective equipment, training, respiratory protection, and the use of fire fighting equipment. Besides Subpart L, Title 29 *CFR* 1910.134 (respiratory protection requirements), Title 29 *CFR* 1910.120 (hazardous waste operations and emergency response requirements), and Title 29 *CFR* 1910.1200 (hazard communication standard) were particular concerns to the fire service. Fire departments were covered by numerous other applicable regulations of the general industry standards (Title 29 *CFR* 1910) and parts of the construction standards (Title 29 *CFR*

1926), in particular Subpart P on excavations. Basically, Title 29 *CFR* 1910 was the fire service's introduction to mandated safety and health requirements. In the years since the adoption of Title 29 *CFR* 1910, these requirements have increased with mandates for the following:

- Hazardous materials mitigation
- Confined space entry
- Increased respiratory protection
- Incident management
- Occupational safety and health
- Health and safety officer
- Infection control, identification, and notification

Because federal OSHA regulations do not apply to all states, it is important for fire service personnel to check with the individual state department of labor concerning the applicable health and safety regulations.

State Occupational Safety and Health Plans

Section 18 of the Occupational Safety and Health Act of 1970 encourages states to develop and operate their own job safety and health programs. These state plans are approved and monitored by OSHA. Currently, the following states and territories have OSHA-approved state plans:

Alaska, Arizona, California, Connecticut, Hawaii, Indiana, Iowa, Kentucky, Maryland, Michigan, Minnesota, Nevada, New Mexico, New York, North Carolina, Oregon, Puerto Rico, South Carolina, Tennessee, Utah, Vermont, Virgin Islands, Virginia, Washington, Wyoming

NOTE: The Connecticut and New York plans cover public sector (state and local government) employment only. Addresses of state OSHA offices are included in Appendix B, Directory of States with Approved Occupational Safety and Health Plans.

 NFPA 1500 and 1521

With such emphasis being placed on employee safety and health, the fire service recognized the need for this process to become a fixture within the operations and management of fire departments. To address the safety needs of the fire service and meet the mandates of the federal government, the National Fire Protection Association (NFPA) published the first edition of NFPA 1500, *Standard on Fire Department Occupational Safety and Health Program* in 1987. This standard provided the minimum requirements for a fire-service-related occupational safety and health program. It specified health and safety guidelines for fire department and emergency services personnel involved in rescue, fire suppression, emergency medical services, hazardous materials operations, special operations, and other related activities. In addition to operational safety requirements, the standard mandated a safety training program, a health and safety committee, and the position of a health and safety officer within the organization. Although in May, 1977, the NFPA had written a standard titled NFPA 1501, *Standard for Fire Department Safety Officer*, NFPA 1500 established the formal requirement for the position.

During the 1992 revision of NFPA 1501, the numbering of this standard was changed to 1521 to be consistent with the numbering of documents covered under the umbrella of NFPA 1500. NFPA 1521 provides the qualifications, authority, and functions for the fire department health and safety officer. NFPA 1500 was the first standard to require the position of health and safety officer, an important step toward the development of the larger and more comprehensive 1521 standard. Since initial development, the requirements for the fire department health and safety officer have increased to the point that this position now has two distinct and vital roles: department health and safety officer and incident safety officer (Figure 1.2).

The two positions, health and safety officer and incident safety officer, perform very different roles within an organization. In many instances, however, the health and safety officer and the incident safety officer may be the same person. In larger departments, additional personnel may divide the responsibilities of the two positions. In either case, assistant safety officers or incident safety officers should be trained to take the place of the primary officer in the event of an absence.

NOTE: At the time this manual was produced, the NFPA was considering eliminating the NFPA

Figure 1.2 Various National Fire Protection Association standards provide guidelines for the health and safety officer and the incident safety officer.

1521 standard. The plan included dividing the information in NFPA 1521 between NFPA 1500 and NFPA 1021, *Standard for Fire Officer Professional Qualifications*. The final decision on this proposal was not available at the press time of this manual.

Health and Safety Officer

The fire department health and safety officer, who reports directly to the fire chief or the chief's designated representative, is responsible for developing and managing the department's risk management plan. This responsibility involves implementing the plan, monitoring the effectiveness of the plan, and communicating the plan to the department members. In addition, the functions of safety education and training, accident prevention, investigation, and records keeping and data collection are delegated to this person. The health and safety officer assists in the design, evaluation, testing, and acceptance of apparatus, equipment, and protective clothing (Figure 1.3). The health and safety officer ensures that safety inspections are made of all department facilities and that violations are corrected. Finally, the areas of health maintenance, wellness programs, infection control, and critical incident stress management are responsibilities of this position.

Figure 1.3 Among the duties of the health and safety officer is to ensure that personal protective equipment is safe, clean, and maintained.

Figure 1.4 The incident safety officer is an important part of the command staff in the incident management system.

Incident Safety Officer

The incident safety officer is part of the command staff within the incident management system (Figure 1.4). This position is responsible for safety at the incident scene including the following:

- Monitoring fireground conditions and advising the incident commander with status reports

- Evaluating hazards and risks

- Acting as a risk manager and ensuring that the department's risk management plan is followed

- Ensuring the establishment of rehabilitation units

- Ensuring the establishment and deployment of rapid intervention crews

- Keeping the incident commander aware of potential dangers

These activities can be part of the health and safety officer's assignment or be assigned to another member of the department who meets the criteria outlined in the NFPA 1521 standard and departmental operational procedures.

 ## Other Safety-Related References

Besides the NFPA standards, other safety-related references have been developed and published by several different organizations, including the American National Standards Institute (ANSI), the National Institute for Science and Technologies (NIST), the American Society for Testing and Materials (ASTM), and the National Institute for Occupational Safety and Health (NIOSH). Many states/provinces have also implemented their own health and safety regulations. See Appendix C, Applicable Codes and Standards, for a list of known applicable codes, standards, and safety references.

Understanding Codes, Laws, Standards, and the Authority Having Jurisdiction

Understanding the difference between guides, codes, regulations, and standards is extremely important for fire and emergency services personnel. Two general procedures are used in the establishment of occupational safety and health laws. The first is through statutes promulgated or made known to the public by legislative action. The second procedure is through codes, regulations, and standards promulgated by agencies with rule-making authority. The latter procedure is, by far, the most common one and is more readily responsive to the need for change. Promulgation through either course of action has the same force and effect of law.

All regulations, rules, case law, and standards are designed with safety in mind. By following them, liability to responders and the responding agency is more easily managed.

Codes
A *code* is a body of law established either by legislative or administrative agencies with rule-making authority. The code is designed to regulate, within the scope of the code, the topic to which it relates. Examples of codes are fire or building codes such as the *Uniform Fire Code*, developed by the International Conference of Building Officials; the *Standard Building Code*, developed by the Southern Building Code Congress International, Inc.; or the *National Building Code of Canada*, developed by the National Research Council of Canada that specify building separation distances, occupancy capacity, or exit requirements.

Regulations
A *regulation* is an authoritative rule dealing with details of procedures or a rule or order having the force of law, issued by an executive authority of government. Regulations usually provide specific application to Acts of legislation. An example of a regulation is the *Code of Federal Regulations*, Title 29 1910.120.

Standards
A *standard* is any rule, principle, or measure established by authority. The term *occupational safety and health standard* under the Occupational Safety and Health Act of 1970 means "a standard which requires conditions, or adoption or use of one or more practices, means, methods, operations, or processes, reasonably necessary or appropriate to provide safe or healthful employment and places of employment." Perhaps the most commonly known standards in the fire service are those developed by the National Fire Protection Association (NFPA).

Guides
A *guide* is an instrument that provides direction or guiding information. Although such guides do not have the force of law, they may be considered as part of what is "reasonable" in a negligence case when determining the standard of care.

Statutory Laws
Statutory laws pertain to civil and criminal matters. Due to the nature of these laws, they may not take effect for many years. These laws regulate such things as water and air pollution through the Clean Water and Air Act, hazardous waste site cleanup through the Superfund Amendments and Reauthorization Act of 1986, and pollution by petroleum products through the Oil Pollution Act of 1990.

Case Law
Case law is usually the result of a legal precedent or a judicial decision. These serve as rules for the future determinations in similar cases. The impact of these decisions affect emergency responders almost immediately because there is usually no implementation period. Case law, if heard at the federal level, can have nationwide effect. Some case law decisions can impact emergency responders even though the individual case did not involve emergency response personnel. An example of a case law decision is *Whirlpool Corporation v. Marshall*, which determined that an employer could not terminate the employment of a worker who refuses to perform an unsafe act that he or she considers unacceptably risky. However, case law is always subject to change and can provide precedence to both sides of an issue.

The phrase *authority having jurisdiction* (AHJ) is defined as "the organization, office, or individual responsible for approving equipment, an installation, or a procedure." The NFPA uses this phrase in its documents in a broad manner because jurisdiction and approval agencies vary, as do their responsibilities.

When standards such as those developed by the NFPA are adopted by ordinance, the authority having jurisdiction has discretion in interpreting and enforcing the consensus standards and can provide equivalent codes as long as the intent of the standard is met. Failure to adopt standards or codes can make the authority having jurisdiction liable for court action. In civil cases, departments are not immune from liability because they did not adopt a consensus standard.

Where public safety is primary, the authority having jurisdiction may be an agency or an individual. Some examples include the following:

- Federal department or agency
- State/provincial department or agency
- Local and regional departments or agencies

- Fire chief
- Fire marshal
- Board of fire commissioners/engineers
- Board of directors
- Chief of the fire prevention bureau, labor department, or health department
- Building official
- Electrical inspector
- Any others having statutory authority

For insurance purposes, the AHJ may include the insurance inspection department, rating bureau, and any other insurance company representative. These last three groups tend to work more for the benefit of insurance companies than for the creation of safety guidelines.

In many circumstances, the property owner or designated agent assumes the role of authority having jurisdiction. In government installations, the commanding officer or department official may be the authority having jurisdiction.

 ## Need for the Fire Department Health and Safety Officer

The fire service has joined with state/provincial and federal agencies along with private industry in an attempt to ensure the safety of fire service personnel. By working jointly on standards committees and product design groups, members of the fire service have proven that safety is a team effort. Through the work of these committees and design groups, the members are helping to reduce firefighter death and injuries; therefore providing longer and healthier retirements for emergency response personnel. This spirit of teamwork and cooperation should also be part of the individual fire department's health and safety programs. The person responsible for creating and maintaining this spirit within the fire department is the health and safety officer.

Attitudes toward safety and health in the fire service are gradually improving. Occupational safety and health requirements have positively changed since the 1970s. Factors that influenced this change include public awareness of the importance of safety and a desire for an improved quality of life on the part of firefighters. Legislation at all levels of government — federal, provincial, state, and local — now regulates components of the fire service occupational safety and health process (Figure 1.5). Among the many areas addressed by fire service occupational safety and health legislation are the following:

- Development of an occupational safety and health program
- Facility safety
- Training and education
- Personal protective equipment
- Respiratory protection
- Infection control
- Health maintenance/wellness
- Hazardous materials right-to-know

Employee safety and health have been legislated because it reflects society's values on the quality of human life. But compliance is not the only reason for them. The following are some of the legal, economic, and ethical factors to consider:

- *Legal responsibilities* — The fire department has the responsibility for providing an occupational safety and health program based upon applicable laws, codes, and standards. Management has legal and ethical obligations to comply with the law. However, federal OSHA regulations apply over local governmental entities in those states and territories that have OSHA-approved state plans. In these states and territories, firefighters are usually covered by federal OSHA regulations. Additional information can be found at the official OSHA web site on the Internet or by writing OSHA directly. These addresses are included at the end of this manual in Appendix B.

- *Ethical factors* — By providing an occupational health and safety program for employees, the fire chief takes a proactive approach to ensuring a safe and healthy work environment. Basically, this approach means making safety a primary

part of the department's policies and procedures. The administration must show that it supports the creation, implementation, and enforcement of the health and safety program.

- *Economic considerations* — From a financial perspective, a proactive occupational safety and health program is a preventive means for protecting the department's assets and protecting against catastrophic loss. The cost of prevention is offset by the savings from lost-time injuries, training of replacement employees, worker's compensation costs, equipment replacement, and possible litigation.

The fire department health and safety officer is the link between management and the line personnel. The safety officer ensures that policies are in place and followed. By stressing the importance of following safety procedures and emphasizing the benefit of personal protection, the safety officer provides a measure of safety in the workplace.

Figure 1.5 The health and safety officer may be assigned the duty of inspecting safety equipment, such as fire extinguishers, within fire department facilities.

◆ Scope of the Fire Department Health and Safety Officer

The occupational safety and health program defines the scope of the health and safety officer's responsibilities within the fire department. Although the fire chief has the ultimate responsibility for the implementation of an occupational safety and health program, the health and safety officer has the assigned task of developing and implementing the safety program if such a program is not already in place. If one is in place, the health and safety officer then manages the program to ensure a safe and healthy work environment.

In addition, the health and safety officer is involved in the development, implementation, and management of the department's written risk management plan. A *risk management plan* establishes goals and objectives to ensure that the risks associated with the operations of the department are identified and managed. The health and safety officer monitors the effectiveness of the plan and communicates the plan through training and education to all members. He will develop an incident risk management plan as part of the incident management system. The health and safety officer uses this plan to monitor safety at emergency incidents. This plan also provides the incident safety officer with a model for monitoring emergency incidents as indicated in Chapter 9.

The responsibilities of the health and safety officer include implementing hiring standards and tests; safety education and training (Figure 1.6); accident prevention and investigation review; equipment and apparatus design, evaluation and acceptance; and station safety (Figure 1.7). Health monitoring, infection control, critical incident stress management, post-incident analysis, incident management safety, and records management and data collection also fall within the health and safety officer's area of authority. Besides coordinating with other department officers to ensure a safe working environment, the health and safety officer may also be called upon to work with the safety sections of other departments within the governmental body.

Even though the health and safety officer reports to the chief of the fire department, he must have good working relationships with all officers and members. These relationships require good communication skills, mutual respect, and compassion on the part of the safety officer. The position is an essential link between management and the firefighter. The health and safety officer also acts as chair of the department's health and safety committee (Figure 1.8). This committee is required by Section 2-6.1 of NFPA 1500. The committee is composed of members of the administration, representatives of member organizations, and individual members in addition to the health and safety officer. The purpose of the committee is to conduct research, develop recommendations, and advise the chief of the department on safety matters. Regularly scheduled meetings must be held every six months. Written meeting minutes must be recorded, retained, and made available for all members of the department. Joint safety and health committees help to improve the communication

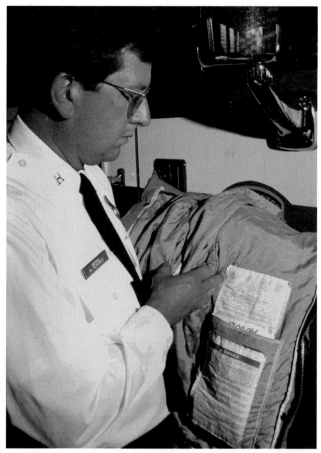

Figure 1.7 The health and safety officer must ensure that personal protective clothing meets the requirements of the appropriate NFPA standard.

Figure 1.6 Safety education and training are elements of the health and safety officer's role.

Figure 1.8 The health and safety officer chairs the department's health and safety committee.

between management and firefighters and strengthen the commitment to the safety and health program.

Qualifications for the Fire Department Safety Officer

Personnel filling the positions of health and safety officer and incident safety officer must have the knowledge, skills, and experience to perform the tasks effectively and efficiently. Qualifications for these positions are specified in the appropriate sections of NFPA 1521.

Health and Safety Officer

NFPA 1521 states that the qualifications for the health and safety officer shall be a Fire Officer I in accordance with NFPA 1021, *Standard for Fire Officer Professional Qualifications*. A jurisdiction may choose to fill this position with a firefighter who does not meet this qualification or with a qualified civilian. Although the authority having jurisdiction may establish an equivalent level of qualifications for the position of health and safety officer, these qualifications may not be lower than the competency standard set by the NFPA standard. The health and safety officer shall be knowledgeable in current applicable laws, codes, and standards pertaining to fire service safety. He shall have knowledge of occupational safety and health hazards inherent to emergency operations and of current principles and techniques of safety management. A working knowledge of current health maintenance and physical fitness issues is a requirement of the position. Another requirement is to have knowledge of infection control practices and procedures as given in NFPA 1581, *Standard on Fire Department Infection Control Program*. Besides these NFPA 1521 requirements, the health and safety officer should have good oral and written communication skills and computer skills. Extensive knowledge of emergency operations, building construction, hazardous materials, fire apparatus design, ergonomics, and incident management are also requirements of the position. The health and safety officer must be able to conduct major research projects, document results, and maintain an effective data management system.

Incident Safety Officer

The position of incident safety officer shall be a fire department officer meeting the requirements of Fire Officer Level I. He shall have the knowledge, skills, and ability to manage an incident scene safely and have knowledge of safety and health hazards involved in emergency operations. Other requirements include knowledge of fire dynamics, building construction, the department's accountability system, and the operation of an emergency scene rehabilitation section.

Authority of the Fire Department Safety Officer

Both the health and safety officer and the incident safety officer need the proper authority to carry out the tasks for which they have responsibility. This authority must come from the fire chief or department head.

Health and Safety Officer

A primary responsibility of the fire chief is to develop and implement an occupational safety and health program. This program ensures the safety and health of all personnel and supports compliance with several NFPA standards and OSHA regulations. The fire chief may appoint a health and safety officer to develop the written occupational safety and health program. The health and safety officer acts with the authority of the fire chief or the management of the fire department. The health and safety officer is responsible for the identification and cause correction of safety and health hazards that affect fire department personnel. He also has the authority to cause the immediate correction of situations that create imminent danger to personnel (Figure 1.9). In situations involving nonimminent hazards, the health and safety officer shall develop actions to correct the situations within the administrative process of the department and bring notice of such hazards to whoever has the ability to correct the problem. The health and safety officer may contribute to the strategic planning of the department but does not have the authority to issue orders or change procedures without the approval of the chief or his designate.

Incident Safety Officer

The incident safety officer's authority as part of the incident command staff is defined in NFPA 1521, 2-5. At an emergency incident where activities are judged to be unsafe or to involve an imminent hazard by the incident safety officer, that person shall have the authority to alter, suspend, or terminate those activities (Figure 1.10). He shall immediately inform the incident commander of any actions taken to correct imminent hazards at the emergency scene. At an emergency incident where the incident safety officer identifies hazards that do not pose an imminent danger, the officer shall take appropriate action through the incident commander to mitigate or eliminate the hazard. If the incident is large enough to warrant more than one incident safety officer, assistants shall be appointed with the necessary authority to perform the same functions.

 Summary

The development and implementation of the fire department occupational safety and health program and the establishment of the health and safety officer position is a critical function of the fire department. Sufficient data exists to justify incorporating this function into the daily operations of a fire department. Due to the significance of the roles and responsibilities of the fire department health and safety officer, this function has been divided into two distinct roles: health and safety officer and incident safety officer. Each function plays a key role for ensuring the success of fire department operations in both nonemergency and emergency settings.

Great strides have been made with regard to firefighter safety and health. Laws, codes, and standards continue to play a profound and pivotal role in the process; however, more work still needs to be done.

Risk management is another component of the occupational safety and health program that has tremendous impact on improving firefighter well-being. The health and safety officer and the incident safety officer must understand and use this process, which will vastly improve operations and reduce the frequency and severity of accidents, injuries, and illnesses.

The establishment and implementation of a health and safety program is not an overwhelming task. With management support, the cooperation of department members, good communications, and teamwork, the program can be established along with the positions of health and safety officer and incident safety officer. The functions and responsibilities of these positions may seem complex but are well within the reach of dedicated individuals. The ultimate goals are the improved health and safety of all members of the department.

Figure 1.9 The health and safety officer has the authority to correct situations that create an imminent danger to personnel such as fighting a fire without proper personal protective equipment in use. *Courtesy of Ron Jeffers.*

Figure 1.10 The incident safety officer advises the incident commander on safety-related issues at emergency incidents.

The Health and Safety
Officer as a Risk Manager

2

This chapter provides information that will assist the reader in meeting the following job performance requirements from NFPA 1521, *Standard for Fire Department Safety Officer,* 1997 edition.

3-1.1 The health and safety officer shall be involved in the development, implementation, and management of the official written risk management plan as specified in Chapter 2 of NFPA 1500, *Standard on Fire Department Occupational Safety and Health Program.*

3-1.1.1 The health and safety officer shall communicate the health and safety aspects of the risk management plan to all members through training and education.

3-1.1.2 The health and safety officer shall make available the written risk management plan to all fire department members.

3-1.2 The health and safety officer shall monitor the effectiveness of the risk management plan and shall ensure the risk management plan is revised annually as it relates to fire fighter health and safety.

3-1.3 The health and safety officer shall develop an incident risk management plan that is implemented into the fire department's incident management system. This risk management plan shall meet the requirements of Chapter 6 of NFPA 1500, *Standard on Fire Department Occupational Safety and Health Program.*

3-3.1 The health and safety officer shall ensure that training in safety procedures relating to all fire department operations and functions is provided to fire department members. Training shall address recommendations arising from the investigation of accidents, injuries, occupational deaths, illnesses, and exposures and the observation of incident scene activities.

3-3.2 The health and safety officer shall cause safety supervision to be provided for training activities, including all live burn exercises. All structural live burn exercises shall be conducted in accordance with NFPA 1403, *Standard on Live Fire Training Evolutions.* The health and safety officer or qualified designee shall be personally involved in preburn inspections of any acquired structures to be utilized for live fire training.

3-3.3 The health and safety officer shall develop and distribute safety and health information for the education of fire department members.

3-4.1 The health and safety officer shall manage an accident prevention program that addresses the items specified in this section. The health and safety officer shall be permitted to delegate the development, direct participation, review, or supervision of this program.

3-4.2 The accident prevention program shall provide instruction in safe work practices for all fire department members. This shall include safe work practices for emergency and nonemergency operations.

3-4.3 The accident prevention program shall address the training and testing of all fire department drivers, including all fire apparatus driver/operators.

3-4.4 The health and safety officer shall periodically survey operations, procedures, equipment, and fire department facilities with regard to maintaining safe working practices and procedures. The health and safety officer shall report any recommendations to the fire chief or the fire chief's designated representative.

3-5.1 The health and safety officer shall develop and implement procedures to ensure that a member(s) suffering a life-threatening occupational injury or illness is provided immediate emergency care and transportation to medical facilities. Those procedures shall also ensure that all occupational injures and illnesses are treated at the most appropriate health care facilities.

3-5.2 The health and safety officer shall investigate, or cause to be investigated, all occupational injuries, illnesses, exposures, and fatalities, or other potential hazardous conditions involving fire department members and all accidents involving fire department vehicles, fire apparatus, equipment, or fire department facilities.

3-5.3 The health and safety officer shall develop corrective recommendations that result from accident investigations. The health and safety officer shall submit such corrective recommendations to the fire chief or the fire chief's designated representative.

3-5.4 The health and safety officer shall develop accident and injury reporting and investigation procedures and shall periodically review these procedures for revision. These accident and injury reporting procedures shall comply with all local, state, and federal requirements.

3-5.5 The health and safety officer shall review the procedures employed during any unusually hazardous operation. Wherever it is determined that incorrect or questionable procedures were employed, the health and safety officer shall submit corrective recommendations to the fire chief or the fire chief's designated representative.

3-6.1 The fire department shall maintain records of all accidents, occupational deaths, injuries, illnesses, and exposures in accordance with Chapter 2 of NFPA 1500, *Standard on Fire Department Occupational Safety and Health Program.* The health and safety officer shall manage the collection and analysis of this information.

3-6.2 The health and safety officer shall identify and analyze safety and health hazards and shall develop corrective actions to deal with those hazards.

3-6.3 The health and safety officer shall ensure that records on the following are maintained as specified in Chapter 2 of NFPA 1500, *Standard on Fire Department Occupational Safety and Health Program*:

(a) Fire Department safety and health standard operating procedures.

(b) Periodic inspection and service testing of apparatus and equipment.

(c) Periodic inspection and testing of personal safety equipment.

(d) Periodic inspection of fire department facilities.

3-6.4 The health and safety officer shall maintain records of all recommendations made and actions taken to implement or correct safety and health hazards or unsafe practices.

3-6.5 The health and safety officer shall maintain records of all measures taken to implement safety and health procedures and accident prevention methods.

3-6.6 The health and safety officer shall issue a report to the fire chief, at least annually, on fire department accidents, occupational injuries, illnesses, deaths, and exposures.

3-7.1 The health and safety officer shall review specifications for new apparatus, equipment, protective clothing, and protective equipment for compliance with applicable safety standards, including the provisions of Chapters 4 and 5 of NFPA 1500, *Standard on Fire Department Occupational Safety and Health Program*.

3-7.2 The health and safety officer shall assist and make recommendations regarding the evaluation of new equipment and its acceptance or approval by the fire department in accordance with the applicable provisions of Chapter 4 of NFPA 1500, *Standard on Fire Department Occupational Safety and Health Program*.

3-7.3 The health and safety officer shall assist and make recommendations regarding the service testing of apparatus and equipment to determine its suitability for continued service and in accordance with Chapter 4 of NFPA 1500, *Standard on Fire Department Occupational Safety and Health Program*.

3-7.4 The health and safety officer shall develop, implement, and maintain a protective clothing and protective equipment program that will meet the requirements of Chapter 5 of NFPA 1500, *Standard on Fire Department Occupational Safety and Health Program*, and provide for the periodic inspection and evaluation of all protective clothing and equipment to determine its suitability for continued service.

3-8.1 The health and safety officer shall ensure that all fire department facilities are inspected in accordance with Chapter 7 of NFPA 1500, *Standard on Fire Department Occupational Safety and Health Program*.

3-8.2 The health and safety officer shall ensure that any safety or health related hazards or code violations are corrected in a prompt and timely manner.

3-9.1 The health and safety officer shall ensure that the fire department complies with the requirements of Chapter 8 of NFPA 1500, *Standard on Fire Department Occupational Safety and Health Program*.

3-9.2 The health and safety officer shall incorporate medical surveillance, wellness programs, physical fitness, nutrition, and injury and illness rehabilitation into the health maintenance program.

3-10.1 The health and safety officer shall be a member of the fire department occupational safety and health committee.

3-10.2 The health and safety officer shall report the recommendations of the fire department occupational safety and health committee to the fire chief or the fire chief's designated representative.

3-10.3 The health and safety officer shall submit recommendations on occupational safety and health to the fire chief or the fire chief's designated representative.

3-10.4 The health and safety officer shall provide information and assistance to officers and fire fighters for surveying their districts, so they will be able to identify and report safety and health hazards that could have adverse effects on fire department operations.

3-10.5 The health and safety officer shall maintain a liaison with staff officers regarding recommended changes in equipment, procedures, and recommended methods to eliminate unsafe practices and reduce existing hazardous conditions.

3-10.6 The health and safety officer shall maintain a liaison with equipment manufacturers, standards-making organizations, regulatory agencies, and safety specialists outside the fire department regarding changes to equipment and procedures and methods to eliminate unsafe practices and reduce existing hazardous conditions.

3-10.7 The health and safety officer shall maintain a liaison with the fire department physician to ensure that needed medical advice and treatment are available to members of the fire department.

3-11.1 The health and safety officer shall ensure that an occupational safety and health committee is established by the fire department.

3-11.2 The health and safety officer shall ensure that the occupational safety and health committee meets the requirements of Chapter 2 of NFPA 1500, *Standard on Fire Department Occupational Safety and Health Program*, and Section 3-9 of this standard.

3-12.1 The health and safety officer shall ensure that the fire department's infection control program meets the requirements of 29 CFR 1910.1030, *Occupational Exposure to Bloodborne Pathogens*, and NFPA 1581, *Standard on Fire Department Infection Control Program*.

3-12.2 The health and safety officer shall maintain a liaison with the person or persons designated as infection control officer to assist in achieving the objectives of the infection control program as specified in NFPA 1581, *Standard on Fire Department Infection Control Program*.

3-12.3 The health and safety officer shall function as the fire department infection control officer if an infection control officer position does not exist in the fire department.

3-13.1 The health and safety officer shall ensure that the fire department establishes a critical incident stress management (CISM) program. The critical incident stress management program shall meet the requirements of Chapter 10 of NFPA 1500, *Standard on Fire Department Occupational Safety and Health Program*.

3-13.2   The health and safety officer shall ensure that the critical incident stress management program is incorporated into the fire department's member assistance program.

3-14.1 The health and safety officer shall develop procedures to ensure that safety and health issues are addressed during postincident analysis.

3-14.2 The health and safety officer shall provide a written report that includes pertinent information about the incident relating to safety and health issues.

3-14.3 The health and safety officer shall include information based upon input from the incident safety officer. This information shall include the incident action plan and the incident safety plan.

3-14.4 The health and safety officer shall include information about issues relating to the use of protective clothing and equipment, personnel accountability system, rehabilitation operations, and other issues affecting the safety and welfare of personnel at the incident scene.

4-1.1 The incident safety officer shall be integrated with the incident management system as a command staff member, as specified in NFPA 1561, *Standard on Fire Department Incident Management System*.

4-1.2 Standard operating procedures shall define criteria for the response or appointment of an incident safety officer. If the incident safety officer is designated by the incident commander, the fire department shall establish criteria for appointment based upon 4-1.1 of this standard.

4-1.3 The incident safety officer and assistant incident safety officer(s) shall be readily identifiable on the incident scene.

4-2.1 The incident safety officer shall monitor conditions, activities, and operations to determine whether they fall within the criteria as defined by the fire department's risk management plan. When the perceived risk(s) is not within these criteria, the incident safety officer shall take action as outlined in Section 2-5.

4-2.2 The incident safety officer shall ensure that the incident commander establishes an incident scene rehabilitation tactical level management unit during emergency operations.

4-2.3 The incident safety officer shall monitor the scene and report the status of conditions, hazards, and risks to the incident commander.

4-2.4 The incident safety officer shall ensure that the fire department's personnel accountability system is being utilized.

4-2.5 The incident commander shall provide the incident safety officer with the incident action plan. The incident safety officer shall provide the incident commander with a risk assessment of incident scene operations.

4-2.6 The incident safety officer shall ensure that established safety zones, collapse zones, hot zone, and other designated hazard areas are communicated to all members present on scene.

4-2.7 The incident safety officer shall evaluate motor vehicle scene traffic hazards and apparatus placement and take appropriate actions to mitigate hazards.

4-2.8 The incident safety officer shall monitor radio transmissions and stay alert to transmission barriers that could result in missed, unclear, or incomplete communication.

4-2.9 The incident safety officer shall communicate to the incident commander the need for assistant incident safety officers due to the need, size, complexity, or duration of the incident.

4-2.10 The incident safety officer shall survey and evaluate the hazards associated with the designation of a landing zone and interface with helicopters.

4-3.1 The incident safety officer shall meet provisions of Section 4-2 of this standard during fire suppression operations.

4-3.2 The incident safety officer shall ensure that a rapid intervention crew meeting the criteria in Chapter 6 of NFPA 1500, *Standard on Fire Department Occupational Safety and Health Program,* is available and ready for deployment.

4-3.3 Where fire has involved a building or buildings, the incident safety officer shall advise the incident commander of hazards, collapse potential, and any fire extension in such building(s).

4-3.4 The incident safety officer shall evaluate visible smoke and fire conditions and advise the incident commander, tactical level management unit officers, and company officers on the potential for flashover, backdraft, blow-up or other fire event that could pose a threat to operating teams.

4-3.5 The incident safety officer shall monitor the accessibility of entry and egress of structures and the effect it has on the safety of members conducting interior operations.

4-4.1 The incident safety officer shall meet provisions of Section 4-2 of this standard during EMS operations.

4-4.2 The incident safety officer shall ensure compliance with the department's infection control plan and NFPA 1581, *Standard on Fire Department Infection Control Program*, during EMS operations.

4-4.3 The incident safety officer shall ensure that incident scene rehabilitation and critical incident stress management are established as needed at EMS operations, especially mass casualty incidents (MCI).

4-5.1 The incident safety officer shall meet provisions of Section 4-2 of this standard during hazardous materials incidents.

4-5.2 The hazardous materials incident safety officer shall meet the requirements of Chapter 4 of NFPA 472, *Standard for Professional Competence of Responders to Hazardous Materials Incidents.*

4-5.3 The incident safety officer shall attend strategic and tactical planning sessions and provide input on risk assessment and member safety.

4-5.4 The incident safety officer shall ensure that a safety briefing, including an incident action plan and an incident safety plan, is developed and made available to all members on the scene.

4-5.5 The incident safety officer shall ensure that hot, warm, decontamination, and other zone designations are clearly marked and communicated to all members.

4-5.6 The incident safety officer shall meet with the incident commander to determine rehabilitation, accountability, or rapid intervention needs. For long-term operations, the incident safety officer shall ensure that food, hygiene facilities, and any other special needs are provided for members.

4-6.1 The incident safety officer shall meet provisions of Section 4-2 of this standard during special operations incidents. The individual that serves as the incident safety officer for special operations incidents shall have the appropriate education, training, and experience in special operations.

4-6.2 The incident safety officer shall attend strategic and tactical planning sessions and provide input on risk assessment and member safety.

4-6.3 The incident safety officer shall ensure that a safety briefing, including an incident action plan and an incident safety plan, is developed and made available to all members on the scene.

4-6.4 The incident safety officer shall meet with the incident commander to determine rehabilitation, accountability, or rapid intervention needs. For long-term operations, the incident safety officer shall ensure that food, hygiene facilities, and any other special needs are provided for members.

4-7.1 Upon notification of a member injury, illness, or exposure, the incident safety officer shall immediately communicate this to the incident commander to ensure that emergency medical care is provided.

4-7.2 The incident safety officer shall initiate the accident investigation procedures required by the fire department.

4-7.3 In the event of a serious injury, fatality, or other potentially harmful occurrence, the incident safety officer shall request assistance from the health and safety officer.

4-8.1 The incident safety officer shall prepare a written report for the postincident analysis that includes pertinent information about the incident relating to safety and health issues. The incident safety officer shall participate in the postincident analysis.

4-8.2 The incident safety officer shall include information about issues relating to the use of protective clothing and equipment, personnel accountability system, rapid intervention crews, rehabilitation operations, and other issues affecting the safety and welfare of members at the incident scene.

Chapter 2

The Health and Safety Officer as a Risk Manager

An idle glance at the roles, responsibilities, and qualifications for the health and safety officer and the incident safety officer outlined in NFPA 1521, *Standard for Fire Department Safety Officer*, can give the impression of an overwhelming challenge. Personnel who fill these positions must be capable of performing multiple skills such as training, investigating, evaluating, analyzing, implementing, and communicating. Each of these skills can be applied to the tasks required of the health and safety officer and incident safety officer. In each task, however, the health and safety officer and incident safety officers are risk managers, applying the elements of risk management to each of the various tasks. To understand how this works, it is necessary to understand the risk management model incorporated in NFPA 1500, *Standard on Fire Department Occupational Safety and Health Program*. The risk management model adopted by the NFPA has been successfully used by general industry for decades. The safety and health components of risk management were incorporated into NFPA 1500 during the 1992 revision process. In the 1997 edition of the standard, they are outlined in Chapter 2 and applied to emergency operations in Chapter 6. The role of the health and safety officer in the risk management process is defined in Chapter 3 of NFPA 1521.

Fire Service Risk Management Plan

The risk management plan described in NFPA 1500 is a process that incorporates several components that can be applied to the operations of a fire department or EMS organization. This plan is not a stagnant document that is developed, described in a printed document, and placed in a manual on a shelf and used occasionally (Figure 2.1). Essentially, a risk management plan serves as documentation that risks have been identified and evaluated and that a reasonable control plan has been implemented and followed. An effective risk management program has a positive impact on the department from the operational, safety, financial, and liability standpoints.

The fire chief or authority having jurisdiction (AHJ) has the ultimate responsibility for the risk management plan (like the occupational safety and health program) but delegates it to the health and safety officer. This risk management plan must be reviewed and revised annually if necessary. Responsibility for the review and revision may be assigned to the health and safety officer, the health and safety committee, or retained by the fire chief. The plan may be a part of the overall community/

Figure 2.1 The health and safety officer uses the risk management plan to minimize the potential risk faced by firefighters while performing their duties. *Courtesy of Chris Mickal.*

jurisdiction plan and may be the responsibility of that organization's risk manager such as the community health and safety manager or loss control manager.

In Chapter 2 of NFPA 1500, the requirements of the risk management plan are simply stated. The fire department shall adopt an official written risk management plan that covers administration, facilities, training, vehicle operations, protective clothing and equipment, operations at emergency incidents, operations at nonemergency incidents, and other related activities. At a minimum, the plan shall include risk identification, risk evaluation, risk control techniques, and risk monitoring. However, in this chapter we are going to add an additional step in the process called "prioritization."

Risk Identification

To identify the risks, the health and safety officer compiles a list of all emergency and nonemergency operations and duties in which the department participates. Ideally, he should take into consideration the worst possible conditions or potential events, including major disasters and multiple events. There are many sources to assist with this identification process. The first, and possibly the most effective, is the department's loss prevention data, which consists of annual fire loss reports by occupancy type, loss value, frequency, etc. Although most departments are too small to rely on their own database for a statistically valid trend, national averages and trends are available from NFPA and the National Fire Academy. It is important to note that national data is not always complete or accurate due to collection inconsistencies and that a time lag of one to two years is required to collect, analyze, and publish it. The health and safety officer should seek input and ideas from department personnel, trade journals, professional associations, and other service providers to identify the potential risks. When using information provided by other fire departments or organizations, the health and safety officer should consider local circumstances when formulating the list of all emergency and nonemergency operations and duties. Other risk identification sources include risk management plans developed by local industry and hazardous substance sites, vulnerability analyses, and Environmental Protection Agency (EPA) plans among others.

Steps in the Development of a Risk Management Plan

1. *Risk identification.* For every aspect of the operation of the fire department at the station, list potential problems. The following are examples of sources of information that may be useful in the process:
   - List of the risks to which members are or may be exposed
   - Records of previous accidents, illnesses, and injuries (both locally and nationally)
   - Facility and apparatus survey/inspection results

2. *Risk evaluation.* Evaluate each item listed in the risk identification process using the following two questions:
   - What is the potential frequency of occurrence?
   - What are the potential severity and expense of its occurrence?

   Use this information to set priorities in the control plan (needs assessment). Some sources of information include the following:
   - Safety audits and inspection reports
   - Prior accident, illness, and injury statistics
   - Application of national data to local circumstances
   - Professional judgment in evaluating risks unique to the jurisdiction

3. *Risk control.* Once the risks are identified and evaluated, determine which control should be implemented and documented. The two primary methods of controlling risk, in order of preference, are the following:
   - Wherever possible, totally eliminate/avoid the risk or the activity that presents the risk. For example, if the risk is falling on ice, then do not allow members to go outside when icy conditions are present.
   - Where it is not possible or practical to avoid or eliminate the risk, take steps to control it. In the previous example, methods of control would be applying sand/salt or wearing of proper footwear.

   Also consider the specific development of safety programs, standard operating procedures, training and inspections as control methods.

4. *Risk management monitoring and follow-up.* Periodically evaluate the selected controls to determine whether they are working satisfactorily. If not, identify and implement new control measures.

Risk Evaluation

Once the health and safety officer identifies the risks, they can be evaluated from both frequency and severity standpoints (Table 2.1). Frequency, referred to by OSHA as "incidence rate," addresses the likelihood of occurrence. Typically, if a particular type of incident such as injuries related to lifting has occurred repeatedly, it will continue to occur until a job hazard or task analysis has been performed to identify the root causes and effective

Table 2.1 Anytown Fire Department Control Measures					
Identification	**Frequency/ Severity**	**Priority**			**Summary of Control Measures**
Strains and sprains	High/medium	High	1.	O	Periodic awareness training for all members
			2.	O	Evaluate function areas to determine location and frequency of occurrence
			3.	O	Based upon outcome of evaluation, conduct a task analysis of identified problems
Stress	Low/high	High	1.	O	Continue health maintenance program
			2.	O	Member participation in physical fitness program
Exposure to fire products	Low/high	Medium	1.	A	Re-evaluate department's philosophy on mandatory SCBA usage
			2.	O	Revise department policy and procedures on mandatory usage
			3.	A	Retraining and education of personnel on chronic effects of inhalation of by-products of combustion
			4.	A	Provide monitoring process of carbon monoxide (CO) levels at fire scenes, especially during overhaul
Vehicle-related Incidents	Medium/high	High	1.	O	Compliance of department with state motor vehicle laws relating to emergency response
			2.	O	Mandatory department-wide EVOC
			3.	O	Monitor individual member's driving record
Terrorism and the workplace	Low/high	Low	1.	O	Provide awareness training for all personnel
			2.	O	Develop policy and procedures as indicated by need
Incident scene safety	Medium/high	High	1.	O	Revise and implement department incident-management system
			2.	O	Revise current policy on mandatory use of full personal protective equipment including SCBA
			3.	O	Evaluate effectiveness of the department's personal accountability system and make needed adjustments
			4.	A	Train all officers in NFA Incident Safety Officer course
Equipment loss	Low/medium	Medium	1.	O	Review annual accident/loss statistics and implement loss-reduction procedures
			2.	A	Develop procedures for review and recommendation for loss prevention based upon significant loss ($1000+)
			3.	O	Maintain department equipment inventory
Facilities and property	Low/high	Medium	1.	A	Review insurance coverage of contents and facilities for adequate coverage due to catastrophe
			2.	O	All new and renovated facilities incorporate life safety and health designs
			3.	O	Conduct routine safety and health inspections of facilities

NOTE: O = Ongoing   A = Action required

control measures have been implemented. In this example, the health and safety officer or health and safety committee must develop and implement guidelines that outline proper lifting techniques, and physical fitness requirements or provide mechanical aids for lifting. See Chapter 3, Physical Fitness and Wellness Considerations, for information on the job task analysis.

Severity addresses the degree of seriousness of the incident and can be measured in a variety of ways such as lost time away from work, cost of damage, cost of and time for repair or replacement of equipment, disruption of service, or legal costs. Refer to Appendix D for the formula for calculating frequency and severity. Incidents of high frequency and high severity must have the highest priority in the risk analysis while those of low frequency and low severity receive the lowest priority. The method for calculating the risk may vary from department to department.

Risk Prioritization

Taken in combination, the results of the frequency and severity determinations help to establish priorities for determining action. Any risk that has a high probability of occurrence, and which will have serious consequences, deserves immediate action and would be considered a high-priority item. Nonserious incidents with a low likelihood of occurrence are a lower priority and can be placed near the bottom of the "action-required" list.

Risk Control Techniques

Once the health and safety officer prioritizes the risks, it is now time for him to apply risk control measures. Several approaches can be taken to risk control, including risk avoidance, risk transfer, and implementation of control measures.

Risk Avoidance

In any situation, the best risk control choice is risk avoidance. Simply put, personnel should avoid the activity that creates the risk. In an emergency services organization, this approach frequently is impractical. Lifting a stretcher presents a serious back injury risk, but personnel cannot avoid this risk and still provide effective service (Figure 2.2). Therefore, training must be provided in the use of safe lifting techniques. However, risk avoidance

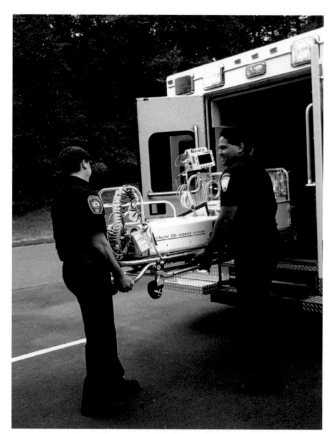

Figure 2.2 Firefighters must be trained in proper lifting techniques to prevent potential back injuries.

could include a policy prohibiting smoking by fire department candidates when they are hired thereby reducing the potential for lung cancer among members.

Risk Transfer

Risk transfer can be accomplished in one of two primary ways: physically transferring the risk to someone else and through the purchase of insurance. For a fire or EMS organization, the transfer of risk may be difficult if not impossible. However, an example of risk transfer would be contracting out the cleanup and disposal of hazardous waste (Figure 2.3). The risks associated with those activities would then transfer to a private contractor who accepts the liability. The purchase of insurance transfers financial risk only. In addition, insurance does nothing to affect the likelihood of occurrence. Buying fire insurance for the station — while highly recommended to protect the assets of the department — does nothing to prevent the station from burning down. Therefore, insurance is no substitute for effective control measures such as installing an automatic sprinkler system.

Figure 2.3 The health and safety officer and the incident safety officer must work with outside contractors in the removal of hazardous waste from incident sites.

Figure 2.4 It is preferable that two firefighters, including one equipped with a portable radio, assist the driver/operator in backing the apparatus.

Control Measures

The most common method used for the management of risk is the adoption of effective control measures (risk reduction). While control measures will not eliminate the risk, they can reduce the likelihood of occurrence or mitigate the severity. Safety, health, and wellness programs; ongoing training and education programs; and well-defined standard operating procedures (SOPs) are all effective control measures.

An example of a situation where control has been very practical is the change in station apparatus bay design and apparatus backing procedures. The risks associated with backing apparatus into station bays are well documented. It is impossible to avoid backing an apparatus into a fire station that is not designed with drive-through apparatus bays. The simplest solutions are improved driver/operator training, painted guidelines on the apparatus bay floors, and a policy that requires a second person to guide the backing operation from the rear of the vehicle (Figure 2.4). A more expensive procedure would be the replacement of older single-door stations with new drive-through stations.

Some typical control measures instituted to control incident scene injuries include use of accountability systems, use of full-protective clothing, a mandatory respiratory protection program, training and education, and health and wellness SOPs. These control measures coupled together make an effective program that promotes safe emergency incident operations.

Monitoring

Once control measures have been implemented, they need to be evaluated to measure their effectiveness. The last step in the process is risk management monitoring. This step ensures that the system is dynamic and facilitates periodic reviews of the entire program. Any problems that occur in the process have to be revised or modified.

The intent of the risk management plan is to develop a strategy for reducing the inherent risks associated with fire department operations. Regardless of the size or type of fire department, every organization should operate within the parameters of a risk management plan. Operation of this plan is a dynamic and aggressive process that the health and safety officer must monitor and revise at least annually or as needed. A sample plan can be found in Appendix A-2-2.1 of the current edition of NFPA 1500.

◆ System Safety Programs

Another form of risk management is the System Safety Program developed by the National Aeronautics and Space Administration (NASA). This program is used by many U.S. government agencies including the Federal Aviation Administration (FAA). The System Safety Program covers the entire spectrum of safety management and assessment, from the design of equipment to the attitude of personnel and the culture of the organization. The system includes the following components:

* Hazard identification and resolution

- Design review
- System modification review and control
- Rules and procedures review
- Equipment design modifications review and control
- Procurement
- Facility and equipment inspections
- Employee and public communications
- Safety training and education
- Emergency response planning, coordination, and training
- Safety data collection and analysis
- Occupational health, wellness, and safety
- Environmental protection
- Interdepartmental and interagency coordination
- Life safety
- Accident and post-incident investigation
- Internal safety and operational audits

Within this framework, there must be documentation, accountability, and verification of results versus requirements. The systems approach provides a logical structure for problem solving, planning, and prevention. It is a tool that can be used with the risk management model to ensure an efficient and effective safety plan for the organization. In addition, a comparison between this list of safety system components and the NFPA 1521 list of roles for the health and safety officer illustrates a similarity between the two.

A detailed description of the System Safety Program can be found in Appendixes 2 and 3 of the FAA System Safety Program, which is available through its web site at www.faa.gov/avr/afs/atos/SIP_system_safety.htm.

 ## Written Safety and Health Policy

The occupational safety and health policy, as required by Chapter 2 of NFPA 1500, establishes the infrastructure for the development and establishment of the department's occupational safety and health program. The purpose of the policy is to clarify for all members their responsibilities in the health and safety process as well as the department's responsibility.

The following is an example of a clear and concise occupational safety and health policy:

It is the policy of the fire department to provide and to operate with the highest possible levels of safety and health for all members. The prevention and reduction of accidents, occupational injuries, and illnesses are goals of the fire department and shall be primary considerations at all times. These concerns for safety and health apply to all members of the fire department and to any other persons who might be involved in fire department activities.

Most fire departments provide other emergency services in addition to fire suppression operations. Based upon the department's risk assessment, members are exposed to a variety of hazards that create a complex set of occupational safety and health issues and concerns. The objective of this policy is to provide the necessary control measures to assist members so that they may perform their assigned tasks safely, efficiently, and effectively. For the policy, like the safety program, to be effective, it must be communicated to all members of the department, reviewed periodically, and applied fairly. It cannot just remain in a binder on a bookshelf in the fire chief's office.

The health and safety officer must develop a series of objectives that meet the goals of the occupational safety and health policy. Objectives may include the establishment of a health and safety program, the creation of a health and safety committee, or the implementation of a health and safety training curriculum.

 ## Health and Safety Program

Each fire department should assign its health and safety officer to develop and implement a safety and health program that meets the requirements, demands, needs, and concerns of the organization. Developing a safety and health program is made easier by using current safety and health laws, codes, and standards as a foundation for program development.

Although these safety standards have requirements for employers, the regulations also contain significant requirements for the employee. Each employee should be required to comply with the requirements of the occupational safety and health standards. One of the principal tasks of the health and safety officer is to manage the department's health and safety program. The health and safety officer is the catalyst who initiates proven safety programs that reduce injury and deaths within fire and emergency medical organizations. Supervisors and members have a responsibility to support and abide by these programs in the interests of improving the work environment. Essential to the success of the program is documented support from the top level of management.

The following are several important reasons for using an occupational safety and health program:

* An ethical obligation for a safety program is evident to prevent injuries, illnesses, and fatalities and to reduce property loss.

* It is a sensible economic investment. The intent of this program is to reduce the frequency and severity of injuries and incidents involving vehicle crashes and property damage, which in turn reduces fire department costs and expenditures to workers' compensation and liability insurance.

* It ensures compliance with applicable laws, codes, and standards.

The safety and health of fire department members are paramount issues for the successful operation of a fire department. There are many components of the occupational safety and health program that ensure a successful process. This process is a comprehensive team effort that requires management support and member participation.

The fire department health and safety officer may also be called upon to administer the appropriate sections of the local jurisdiction's health and safety plan. This plan may include sections on workers' compensation, workplace violence, smoking, and drug and alcohol abuse. If the local jurisdiction does not have a policy, then these issues need to be included in the fire department's health and safety plan.

 ## Occupational Safety and Health Committee

One of the instrumental and valuable components of a fire department's safety and health program is the development, implementation, and operation of an occupational safety and health committee. This committee is a centralized group within a fire department that functions as a clearinghouse for activities, problems, and issues relating to firefighter safety and health.

The issues that confront this committee are unlimited. Ensuring firefighter safety and health are constantly changing processes based upon the needs of the organization and the commitment to safety by the department. Other factors that can impact the successful operation of this committee are the level of activity of the committee and the examining and managing of current safety and health issues within the department. The goals of the occupational safety and health committee are to develop and recommend solutions to resolve conflicts. One issue that is not an objective of this committee is disciplinary action. The activities and issues that are addressed must be within the scope of the committee. NFPA 1500 defines the criteria for establishing and using an occupational safety and health committee. The requirements are as follows:

* The fire department needs to establish an occupational safety and health committee that will serve in an advisory capacity to the fire chief.

* The intent of the occupational safety and health committee is to conduct research, develop recommendations, and study and review matters pertaining to occupational safety and health within the fire department (Figure 2.5).

* The occupational safety and health committee must hold regularly scheduled meetings and special meetings as deemed necessary. Regular meetings must be held at least once every six months. Written minutes of each meeting must be maintained and distributed to all members of the fire department.

The health and safety officer is responsible for directing the operations of the occupational safety and health committee; however, other members of

Figure 2.5 The health and safety committee determines the appropriate level of personal protective equipment needed for the department based on the types of hazards it may encounter.

the committee will vary from department to department depending on size and type. An effective health and safety committee usually consists of three members from management, three members from the member organization, and the health and safety officer as a nonvoting member. Additional members may be appointed including the fire department physician. Committees consisting of more than six members can become unwieldy and ineffective.

Regardless of the size and type of the fire department, every fire department should employ an occupational safety and health committee. A career fire department uses members from each shift, each battalion, each division, or other combination depending on the size of the department.

As with any organization, the key for the success of the organization's safety and health committee is time management. With training mandates as well as other time requirements, the key to participation is efficiency and effectiveness. An organization must emphasize the importance of good time management to make the process work.

 ## The Health and Safety Officer and Incident Safety Officer Roles in Risk Management

Roles and Responsibilities of the Health and Safety Officer

NFPA 1521 is very specific in defining the roles and responsibilities of the health and safety officer. As a minimum standard, it includes (but is not limited to) the following tasks:

- Ensure safety training and education
- Manage the accident or loss prevention program
- Investigate the accident or incident
- Maintain records management and data analysis
- Review equipment specifications and assist in acceptance testing
- Ensure compliance with codes regarding facility safety
- Comply with health maintenance requirements
- Serve as internal and external liaisons
- Act as infection control officer
- Develop a critical incident stress management program
- Conduct a postincident analysis
- Address workplace violence

Ensure Safety Training and Education

The health and safety officer is responsible for ensuring safety training and education for all members of the department (Figure 2.6). This training is related to all types of operations and functions with which the department is likely to be involved. Information collected from accidents, injuries, occupational deaths, job-related illnesses, exposures, and postincident analysis must also be translated into educational material for all members. The health and safety officer is also responsible for proper safety supervision of all fire department training activities including live burn exercises. This role will be expanded upon in

Figure 2.6 It is the health and safety officer's duty to explain the health and safety policy to all department personnel.

Chapter 3, Physical Fitness and Wellness Considerations, and Chapter 4, The Health and Safety Officer's Responsibilities in the Training Process.

Manage the Accident or Loss Prevention Program

A very important part of the health and safety officer's position is managing the accident or loss prevention program. The health and safety officer is ultimately responsible for the development, review, and supervision of the program but may delegate this task. As part of this responsibility, the health and safety officer is required to periodically review operations, procedures, equipment, and facilities and recommend any changes in work practices and procedures to the authority having jurisdiction (Figure 2.7). The accident or loss prevention program includes instruction of all personnel in safe work practices both in emergency and nonemergency operations (Figure 2.8). Fundamental to this program is an effective fire department driver/operator training and testing policy that covers both apparatus and staff vehicle operators.

Investigate the Accident or Incident

The health and safety officer is responsible for investigating incidents that result in hazardous conditions, injuries, illnesses, exposures, and fatalities involving fire department members. He also investigates incidents involving fire department apparatus, vehicles, facilities, and equipment. The health and safety officer is responsible for reviewing the procedures in use at the time of the incident and developing corrective procedures (Figure 2.9).

Figure 2.7 The health and safety officer is responsible for ensuring that warning labels are placed on all apparatus.

Figure 2.8 Safety warning labels include information on the wearing of seat belts, the maximum height of the apparatus, and additional operating information.

Figure 2.9 The health and safety officer investigates any injury or property loss incident involving fire department personnel or equipment. *Courtesy of Ron Jeffers.*

The immediate responsibility following an injury-related incident is to ensure that transportation and medical treatment are provided for the injured member. The health and safety officer is responsible for the development and implementation of the procedures that ensure this care at the most appropriate health care facility.

If necessary, the health and safety officer works closely with law enforcement agencies to ensure complete and accurate reporting in the event of litigation or liability. This information will also be part of the postincident analysis.

The incident review shall include recommendations to the fire chief or AHJ for corrective action to prevent future injury or property loss incidents. Procedures for investigating and reporting these types of incidents shall be in place and reviewed periodically and shall be in compliance with federal, state, and local requirements. Guidelines for investigating incidents are provided in Chapter 4, Health and Safety Officer's Responsibilities in the Training Process.

Maintain Records Management and Data Analysis

Good records management and data analysis provide justification for corrective action. NFPA 1521 requires that departments maintain records on all accidents, occupational deaths, injuries, illnesses, and exposures. It is the duty of the health and safety officer to collect and analyze this information and recommend corrective action (Figure 2.10). The health and safety officer is responsible for the following tasks:

- Health and safety SOPs

- Maintenance of records pertaining to periodic inspections

- Safety testing of department apparatus and equipment

- Inspection and testing of in-service personal protective equipment

- Inspection of department facilities

When recommended safety corrections are made, the health and safety officer keeps records of those corrections in addition to records of the implementation of safety and health procedures and accident prevention policies.

Figure 2.10 The health and safety officer is responsible for collecting, analyzing, and maintaining safety-related data for the department.

The health and safety officer is required to develop an annual safety and health report that is given to the fire chief and communicated to all personnel. This report includes records relating to department accidents, occupational injuries, illnesses, deaths, and hazardous exposures. The report provides a comparison between the department's records and national fire and industry safety and health data. This report helps to determine trends that may be developing. It should be remembered that national data is not always complete and may not be dependable.

Compliance for data management must meet both NFPA 1500 and OSHA regulations. NFPA 1500 requires that the department develop and maintain a data collection system. Permanent records of all job-related accidents, injuries, illnesses, exposures to infectious agents and communicable diseases, or fatalities must be collected in the system. The federal government requires those states that have adopted OSHA regulations or state-approved plans to collect data on all job-related injuries, illnesses, and fatalities. This requirement is regulated in Title 29 *CFR* 1904 and 1952, which provide the record keeping guidelines. A log and summary of all recordable occupational injuries and illnesses must be kept and provided to the Department of Labor and any employee, former employee, or their representative. The Department of Labor/OSHA forms, included in Appendix E of this manual, are Forms 300 and 301. Jurisdictions not covered by OSHA or an approved state plan should consult their state/provincial Department of Labor or Workers' Compensation Board.

Review Equipment Specifications and Assist in Acceptance Testing

In conjunction with the staff officer responsible for the design and purchase of fire apparatus and equipment, the health and safety officer shall review all specifications for new apparatus, equipment, and personal protective clothing and equipment. Specifications must comply with all OSHA, NFPA, and Department of Transportation (DOT) standards and regulations. Upon receipt of new apparatus and equipment, the health and safety officer shall assist in the acceptance testing of the equipment and make recommendations as necessary to ensure compliance with applicable codes, standards, and regulations. The health and safety officer shall be involved in the periodic service testing of fire department equipment including (but not limited to) hose tests, ladder tests, safety and rescue equipment tests, apparatus pump and aerial device tests, and power extrication equipment tests (Figure 2.11).

The health and safety officer is responsible for developing, implementing, and maintaining the protective clothing and equipment program. This responsibility involves writing specifications, performing acceptance tests, performing in-service inspections, implementing a cleaning and repair program, and determining replacement criteria (Figure 2.12). More information on protective clothing can be found later in this manual. See Chapter 6, Equipment Safety and Maintenance; Chapter 7, Personal Protective Equipment; and Chapter 8, Emergency Response and Apparatus Safety.

Ensure Compliance

The health and safety officer shall inspect all fire department facilities for building, safety, and health code violations and shall report all violations, make recommendations for corrections, and ensure compliance with the recommendations (Figure 2.13). Compliance shall be based on NFPA 1500; local building, fire, and health codes; OSHA regulations; National Electrical Code®; and life safety codes as applicable. See Chapter 5, Facilities Safety, for additional information on code compliance. Also, the IFSTA **Inspection and Code Enforcement** manual has additional information regarding building, fire, and electrical requirements.

Figure 2.11 Annual pump tests ensure that all pumping apparatus meet minimum NFPA standards.

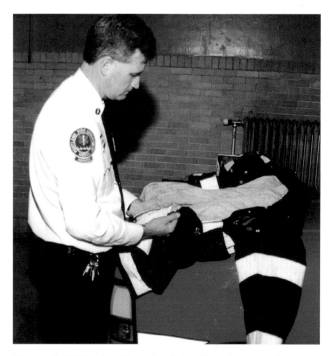

Figure 2.12 The health and safety officer inspects the condition of all personal protective equipment to ensure the safety of firefighters. *Courtesy of Mike Nixon.*

Figure 2.13 Facilities inspections ensure a safe working environment.

Comply with Health Maintenance Requirements

The health and safety officer shall ensure that the fire department complies with the requirements for heath maintenance found in NFPA 1500. Chapter 8 of NFPA 1500 covers medical requirements for hiring firefighters, physical performance and fitness requirements, infection control, individual medical data, and postinjury or illness rehabilitation. It also contains the requirement for a fire department physician. The health maintenance program must also include the components of medical surveillance, physical fitness, wellness, nutrition, and rehabilitation (Figure 2.14). Chapter 3, Physical Fitness and Wellness Considerations, goes into detail on the issues of health maintenance and physical fitness.

Serve as Internal and External Liaisons

The health and safety officer serves as an internal liaison between the department administration and members of the department. He also serves as an external liaison between the department and equipment manufacturers and outside agencies involved in safety issues.

Another important liaison position involves the fire department physician. This link ensures that the medical needs of the department are being met and that the department is aware of any trends that may be developing.

Internal liaison. As the internal liaison, the health and safety officer serves as a member of the department's occupational health and safety committee and reports committee recommendations to the fire chief. This responsibility also includes working with department staff officers in charge of training, equipment purchase and maintenance, and policy development to ensure that unsafe practices and designs are eliminated. The health and safety officer also provides assistance to line personnel in identifying potential hazards within their response areas. This assistance may involve working with fire prevention and inspection personnel, building inspectors, plans review officers, and the hazardous materials officer and team.

External liaison. As the external liaison, the health and safety officer meets and works with equipment manufacturers and vendors to ensure

Figure 2.14 The health and safety officer is responsible for managing the department's health maintenance program including the physical fitness component.

that the equipment acquired by the department meets all standards and specifications. It is also important for the health and safety officer to develop an alliance with members of standards committees and regulatory agencies along with consultants and safety specialists in both the public and private sectors. This alliance can be accomplished by attending fire and safety conferences and manufacturer-sponsored meetings and by serving on standards committees and participating in and networking with local health and safety councils and risk management associations.

Act as Infection Control Officer

The health and safety officer shall act as the department's infection control officer if that position does not already exist (Figure 2.15). If it does, then it is important to maintain a close liaison with that officer to ensure that the requirements of NFPA 1581, *Standard on Fire Department Infection Control Program,* and OSHA Title 29 *CFR* 1910.1030, *Occupational Exposure to Bloodborne Pathogens,* are met. A familiarity with the reporting procedures for the Centers for Disease Control (CDC) is also necessary.

Develop a Critical Incident Stress Management Program

In accordance with Chapter 10 of NFPA 1500, the health and safety officer shall ensure that a critical incident stress management program is developed in the form of a written policy and that this policy is implemented (Figure 2.16). The program will be part of the fire department employee assistance program and available to both members and their families. See Chapter 3, Physical Fitness and Wellness Considerations, for more information on this program.

Conduct a Postincident Analysis

The postincident analysis is an opportunity for the health and safety office to identify, analyze, and

Figure 2.15 In some jurisdictions, the health and safety officer is also the department's infection control officer and is responsible for instructing personnel in the proper use of infection control devices such as the HEPA filter mask.

Figure 2.16 Following incidents that expose department personnel to high levels of stress, the health and safety officer is responsible for the implementation of the critical stress management program.

correct any problems or deficiencies discovered during an incident. The health and safety officer must ensure that safety and health issues are included in this analysis. The health and safety officer's analysis shall include information gathered by the incident safety officer(s), including both the incident action plan and the incident safety officer's safety plan. Additional information pertaining to the use of protective clothing, the personnel accountability system, rehabilitation operations, hazardous conditions, and any other issues relating to the safety of personnel at the incident shall be included in the final written analysis.

Address Workplace Violence

An issue not directly addressed in NFPA 1521, which would fall under the general responsibility of the health and safety officer, is workplace violence (Figure 2.17). The health and safety officer should perform the following tasks:

- Include this type of situation in the health and safety program.

- Monitor conditions and morale of department personnel.

- Train supervisors in the various levels and symptoms of stress and appropriate intervention techniques.

- Recommend appropriate employee assistance programs to help mitigate any potential problems.

Figure 2.17 The health and safety officer must keep up to date on the most current safety-related issues such as studies on workplace violence.

Roles and Responsibilities of the Incident Safety Officer

The effective management of operational risk is the function of the incident commander. The incident commander has the responsibility for ensuring personnel safety at the emergency incident scene. This responsibility is shared with the incident safety officer as outlined in NFPA 1521 and NFPA 1561, *Standard on Fire Department Incident Management System.*

The incident safety officer plays a vital role in the Incident Management System (IMS). The incident safety officer must have the authority from the chief of the department to immediately suspend, terminate, or alter any operation that jeopardizes the safety of department personnel. Command must be immediately informed of any operational changes. The incident safety officer shall also ensure that the health and welfare of members are maintained through the rehabilitation process at an emergency incident, especially during extended emergency operations.

The fire department is required to have a policy in place that ensures the presence of an incident safety officer on scene. Either the department health and safety officer is dispatched to the incident or the incident commander has the authority to appoint a qualified person to fill the position. This individual must meet the qualifications of NFPA 1021, *Standard for Fire Officer Professional Qualifications*, Level I. The person designated as incident safety officer must be readily identifiable by vest, helmet, coat, or armband. Larger fire departments may have enough personnel to establish both the health and safety officer and the incident safety officer positions. Smaller departments may assign the incident safety officer duties to other staff officers or company officers. In any case, the NFPA qualifications must be met.

Incident scene safety as outlined in NFPA 1521 applies to all incidents including (but not limited to) fire suppression operations, emergency medical services operations, hazardous materials operations, special operations, and wildland fire incidents (Figure 2.18). It is the responsibility of the incident safety officer to monitor all conditions, activities, and operations to ensure as safe an environment as possible. This responsibility includes

Figure 2.18 It is the responsibility of the incident safety officer to monitor all emergency incidents, including EMS, and report safety-related issues to the incident commander. *Courtesy of Ron Jeffers.*

reporting to the incident commander on conditions, hazards, and risks; establishing a rehabilitation unit; and monitoring the personnel accountability system. It also includes establishing hazardous materials control zones, collapse zones, and safety zones as required by the circumstances. The incident safety officer monitors radio communications for possible barriers to transmission, observes motor vehicle traffic and apparatus placement for potential risks, and evaluates any hazards that might be associated with the establishment of a landing zone for helicopter operations. In all situations, the incident safety officer keeps the incident commander informed of potential hazards and risks at the incident. See Chapter 9, Incident Scene Safety, for more information on incident scene safety.

In the event of an injury, illness, or fatality involving a member of the department during the incident, the incident safety officer shall immediately notify the incident commander and the health and safety officer and begin accident investigation procedures. In the absence of the health and safety officer, the incident safety officer shall follow the accident procedures outlined for that officer.

Following the incident, the incident safety officer shall prepare a written report including all information about the incident involving health and safety issues. This information should include issues relating to protective clothing, accountability, rapid intervention, rehabilitation, and any other safety-related information. The incident safety officer shall participate in the post-incident analysis critique.

Fire Suppression

During fire suppression operations, the incident safety officer advises the incident commander on potential hazards such as building collapse and fire extension. The incident safety officer also monitors the accessibility for entry into and exit from the structure and ensures that a rapid intervention crew is available and ready for deployment in the event of an emergency (Figure 2.19). The incident safety officer is also responsible for keeping all levels of tactical command advised of potential flashovers, backdrafts, and blowups by monitoring visible smoke and fire conditions around the structure.

Emergency Medical Services

The incident safety officer shall ensure compliance with NFPA 1581 and the fire department's infection control plan during all incidents. When warranted, the incident safety officer shall establish rehabilitation services for responding personnel. If the incident involves mass casualties, the incident safety officer implements the critical incident stress management program of the department.

Hazardous Materials Incidents

At hazardous materials incidents, the incident safety officer works closely with the incident commander, operations officer, hazardous materials safety officer, and the hazardous materials team to protect department personnel (Figure 2.20). In order to ensure this protection, it is necessary that the incident safety officer meet the requirements of NFPA 472, *Standard for Professional Competence of Responders to Hazardous Materials Incidents;* attend both strategic and tactical pre-incident planning sessions; and provide a safety briefing at the incident. During the incident, the person in this position is responsible for ensuring that all elements of the Incident Management System are provided, including rehabilitation, accountability, and rapid intervention. It should be understood that if the incident escalates to the Branch Level of the IMS, the duties of the incident safety officer and the hazardous materials safety officer may be delegated to the same person or to two different officers. OSHA Title 29 *CFR* 1910.120 provides detailed guidelines for the hazardous materials safety officer.

Figure 2.19 During suppression operations, the incident safety officer advises the incident commander on potential hazards including building collapse and fire extension.

Figure 2.20 At hazardous materials incidents, the incident safety officer works with the incident commander, operations officer, hazardous materials safety officer, and the hazardous materials team to ensure the safety of all personnel.

Special Operations

Special operations require that the incident safety officer have the appropriate training, education, and experience to properly assist the incident commander. Incidents of this type include but are not limited to water rescue, building collapse, confined space operations, and aircraft-related incidents. The duties of the incident safety officer, though specific to the type of incident, are similar to those listed for hazardous materials incidents.

Wildland Fire Incidents

Although NFPA 1521 does not specifically address the duties of the incident safety officer at wildland fire incidents, the general incident scene safety and special operations requirements in this standard may be applied to wildland incidents (Figure 2.21). In general, the incident safety officer assesses hazardous and unsafe situations and develops measures for ensuring personnel safety, acts as personnel accountability officer for the incident, and establishes rehabilitation services for response personnel. The incident safety officer has emergency authority to stop and/or prevent unsafe acts. This individual must have knowledge of wildland fire fighting tactics and be able to assess wildland fire indicators. Training material is available through IFSTA and the National Wildland Fire Coordinating Group (NWCG) in addition to national and state forestry departments.

Figure 2.21 Wildland and forestry operations pose special safety concerns for the incident safety officer and the incident commander. *Courtesy of Tony Bacon.*

 Summary

An understanding of the concepts of risk management and of system safety are essential to the positions of health and safety officer and incident safety officer. These concepts are the basis for the majority of the roles and responsibilities of both positions. The ability to identify, analyze, and correct are the keys to developing an effective safety program. This program is managed by the health and safety officer and implemented at the incident scene by the incident safety officer.

3

Physical Fitness and Wellness Considerations

This chapter provides information that will assist the reader in meeting the following job performance requirements from NFPA 1521, *Standard for Fire Department Safety Officer,* 1997 edition.

3-3.3 The health and safety officer shall develop and distribute safety and health information for the education of fire department members.

3-6.1 The fire department shall maintain records of all accidents, occupational deaths, injuries, illnesses, and exposures in accordance with Chapter 2 of NFPA 1500, *Standard on Fire Department Occupational Safety and Health Program.* The health and safety officer shall manage the collection and analysis of this information.

3-6.5 The health and safety officer shall maintain records of all measures taken to implement safety and health procedures and accident prevention methods.

3-6.6 The health and safety officer shall issue a report to the fire chief, at least annually, on fire department accidents, occupational injuries, illnesses, deaths, and exposures.

3-9.1 The health and safety officer shall ensure that the fire department complies with the requirements of Chapter 8 of NFPA 1500, *Standard on Fire Department Occupational Safety and Health Program.*

3-9.2 The health and safety officer shall incorporate medical surveillance, wellness programs, physical fitness, nutrition, and injury and illness rehabilitation into the health maintenance program.

3-10.7 The health and safety officer shall maintain a liaison with the fire department physician to ensure that needed medical advice and treatment are available to members of the fire department.

3-12.1 The health and safety officer shall ensure that the fire department's infection control program meets the requirements of 29 CFR 1910.1030, *Occupational Exposures to Bloodborne Pathogens,* and NFPA 1581, *Standard on Fire Department Infection Control Program.*

3-12.2 The health and safety officer shall maintain a liaison with the person or persons designated as infection control officer to assist in achieving the objectives of the infection control program as specified in NFPA 1581, *Standard on Fire Department Infection Control Program.*

3-12.3 The health and safety officer shall function as the fire department infection control officer if an infection control officer position does not exist in the fire department.

3-13.1 The health and safety officer shall ensure that the fire department establishes a critical incident stress management (CISM) program. The critical incident stress management program shall meet the requirements of Chapter 10 of NFPA 1500, *Standard on Fire Department Occupational Safety and Health Program.*

3-13.2 The health and safety officer shall ensure that the critical incident stress management program is incorporated into the fire department's member assistance program.

4-2.2 The incident safety officer shall ensure that the incident commander establishes an incident scene rehabilitation tactical level management unit during emergency operations.

4-4.2 The incident safety officer shall ensure compliance with the department's infection control plan and NFPA 1581, *Standard on Fire Department Infection Control Program,* during EMS operations.

4-4.3 The incident safety officer shall ensure that incident scene rehabilitation and critical incident stress management are established as needed at EMS operations, especially mass casualty incidents (MCI).

4-5.6 The incident safety officer shall meet with the incident commander to determine rehabilitation, accountability, or rapid intervention needs. For long-term operations, the incident safety officer shall ensure that food, hygiene facilities, and any other special needs are provided for members.

Physical Fitness and Wellness Considerations

Like the post-World War II generations of Americans, firefighters have become increasingly concerned about their physical fitness. Gymnasiums, once the exclusive domain of men, have become health clubs frequented by both men and women of all ages. Products ranging from weight-loss pills and nutritional supplements to abdominal muscle strengtheners and exercise machines are advertised in the media and line store shelves. Fitness and nutrition magazines target this high-dollar market with information on all aspects of health and wellness. The federal government, through the Centers for Disease Control (CDC), has funded studies on the health and wellness of the American people and actively attempted to change some of the habits that are harmful to people. In 1997, the International Association of Fire Fighters (IAFF) and the International Association of Fire Chiefs (IAFC) established the Fire Service Joint Labor Management Wellness/Fitness Initiative. This initiative is intended to improve the quality of life of all firefighters.

Paradoxically, 54 percent of the American adult public is overweight and prone to heart disease, high blood pressure, stroke, and cancer related to improper diets. Eating disorders, obesity, bulimia, and anorexia are not limited to a particular age group or gender. Children are being diagnosed with problems related to weight and nutrition. Eating disorders in men have more than doubled in the last 10 years according to the CDC. Smoking and the use of other tobacco products continue to be popular in all age groups. Smoking is a factor in heart disease, which is the leading cause of death in North America and has resulted in national and state controls on secondhand smoke and cigarette advertising.

The fire service has not been immune to the problems associated with poor physical fitness, poor nutrition, and the use of tobacco products. Since the 1970s, the fire service has become increasingly aware of the issues surrounding health and wellness and has attempted to implement change. This chapter addresses some of the programs and policies designed to improve the health and wellness of firefighters.

Understanding Physical Fitness

To understand the need for physical fitness in the fire service, the health and safety officer should know how physical fitness programs have developed, what the results of such programs should be, and how to develop a task analysis of the jobs performed by a firefighter. In addition, knowledge of the fire department physician's duties and responsibilities is as important as the knowledge of the performance requirements for firefighters in their daily tasks. Other areas that the health and safety officer must be familiar with are the requirements for annual physical fitness evaluations, medical evaluations and examinations, and the keeping of medical records on all personnel. These topics are discussed in the following paragraphs.

History

Fire service concern for the health of firefighters was initially directed at protecting them from hazards associated with fireground operations. This concern included the design and purchase of improved protective clothing and the purchase and use of self-contained breathing apparatus (SCBA). Household items that produced hazardous by-products when burned have increased the use of SCBAs by today's firefighters. The use of SCBA added physical strain to firefighters and increased stress levels. Increased injuries, cardiovascular disease, breathing problems, and loss of stamina during emergency operations indicated a need for improved physical fitness (Figure 3.1).

By the 1980s, the fire service began to see the value of implementing mandatory physical fitness programs. Some of these were based on the then-popular Royal Canadian Air Force Plan 5BX for Men (1962). This plan, which required no equipment other than an exercise mat, took only 12 minutes to complete. It was intended to improve the cardiopulmonary capacity of the participant and to firm the muscles. Although a step in the right direction, the 5BX plan had its problems. It was gender specific (discriminating toward women), did not address work-related tasks, and resulted in injuries to the participants. In one fire department, injuries during the physical fitness program outnumbered those sustained in any other activity.

A suitable replacement program would have to provide solutions not only for the stresses of structural fire fighting but for all the tasks performed by firefighters. According to the CDC, the overall results of a good physical fitness program should include the following:

- Reducing the risk of dying prematurely from all causes
- Reducing the risk of dying prematurely from heart disease
- Reducing the risk of developing adult-onset diabetes
- Reducing the risk of developing high blood pressure
- Reducing high blood pressure in people who already have it
- Reducing the risk of developing colon cancer
- Reducing depression and anxiety
- Controlling weight
- Building healthy bones, muscles, and joints
- Promoting psychological well-being
- Improving sleep habits

In order to develop a comprehensive, holistic physical fitness program, each fire department health and safety officer must prepare an analysis of the tasks performed by its members. This analysis is used as the basis for developing the physical fitness program and for establishing the hiring criteria for new personnel.

Task Analysis

Preparing the task analysis and developing a physical fitness program should be the responsibility of the health and safety committee or a physical fitness subcommittee. Members of the subcommittee should include the health and safety officer, members of the department administration, representatives of the member

Figure 3.1 Emergency operations place excessive physical stress on firefighters. *Courtesy of Ron Jeffers.*

organization, company personnel, the fire department physician and a qualified/certified exercise physiologist. The exercise physiologist provides the professional knowledge necessary to analyze the tasks in terms of physical exertion and recommends the appropriate test criteria.

The first step in the task analysis is for members of the committee to develop a list of the basic services that the fire department provides. These services should include (but are not limited to) structural fire fighting, wildland fire fighting, medical services, light and heavy rescue operations, hazardous materials responses, training functions, and building inspections and surveys (Figure 3.2). Next, the committee members determine the types of tools and equipment that each service activity requires. The tools and equipment, such as hose and nozzles, pike poles, ground ladders, and axes, indicate the types of physical tasks that each activity requires. These physical tasks may include the following:

Figure 3.2 Before creating a physical fitness program for the department, the health and safety officer must develop a task analysis based on the types of tasks performed by firefighters.

• Lifting an inert weight such as an unconscious victim

• Pulling a hose, charged and empty

• Operating a hoseline and nozzle

• Climbing a ladder

• Climbing a flight of stairs in full protective clothing with SCBA

• Entering a confined space with hand tools

• Pulling a ceiling with a pike pole

• Swinging an axe

Once the list of tasks is complete, it can be categorized into general groups of similar activities such as lifting, pulling, climbing, etc. A suitable physical fitness program can be devised to meet the needs of each task.

In 1999, the International Association of Fire Fighters and the International Association of Fire Chiefs introduced a joint initiative on candidate testing. Due to the wide range of preemployment tests in use in North America and the desire to ensure that candidates can perform the tasks required of professional firefighters, the two organizations developed the Candidate Physical Ability Test (CPAT). The test consists of eight separate events along a predetermined sequence or path (Figure 3.3). The candidate must complete all the events within a maximum total time of 10 minutes and 20 seconds. This is a pass/fail test. The eight events include the following:

• Stair climb

• Hose drag

• Equipment carry

• Ladder raise and extension

• Forcible entry

• Victim search

• Victim rescue

• Ceiling breach and pull

The candidate wears a 50-pound (22.68 kg) weighted vest to simulate the weight of a self-contained breathing apparatus and protective clothing. Throughout the test, the candidate wears long trousers, footwear with no open toe or heel, a hard hat

Figure 3.3 The Candidate Physical Ability Test (CPAT) includes eight separate events that simulate fire fighting tasks.

Figure 3.4 The health and safety officer must work closely with the fire department physician to ensure that all personnel are medically fit to meet the demands of their jobs.

with chin strap, and work gloves. Loose or restrictive jewelry and watches may not be worn.

The CPAT provides a content-valid pass/fail test based on the tasks that firefighters perform. It avoids any difficulties that might be associated with testing candidates on skills that require firefighter academy training. A copy of the test can be obtained from the IAFF or IAFC.

Fire Department Physician

In order to develop and implement an effective physical fitness and health maintenance program, proper management and direction must be used. The selection and appointment of a fire department physician is a crucial element. The requirements for the fire department physician are found in Chapter 8, Section 6 of NFPA 1500 and Chapter 2 of NFPA 1582, *Standard on Medical Requirements for Fire Fighters and Information for*

Fire Department Physicians, 2000 edition. See Appendix B of NFPA 1582 for a comprehensive list of these requirements. These requirements include the following:

- The fire department shall officially designate a physician who is responsible for advising, steering, and counseling the members regarding their health, fitness, and suitability for various duties.

- The fire department physician shall provide medical guidance regarding the management of the occupational safety and health program.

- The fire department physician shall be a licensed medical doctor or osteopathic physician who is qualified to provide professional expertise in occupational safety and health relating to emergency services.

- The fire department physician shall be available for consultation and for providing expert services on an emergency basis.

In order to provide an NFPA-compliant medical certification program, the fire department physician must have a clear understanding of the services the department provides (Figure 3.4). The fire department work frequency combined with the task analysis help in providing this understanding. It may be necessary to provide training for the fire department physician in the tasks, operating procedures, and evolutions inherent in fire fighting. Participation on the health and safety committee also helps in his assimilation into the fire service.

Performance Requirements

The task analysis is the basis for determining the level of physical ability for new candidates as well as for current firefighters. A validated preemployment physical ability test is mandated in Chapter 8, Section 2 of NFPA 1500. This section requires that the fire department develop physical performance standards for all personnel who engage in emergency operations. New candidates must also meet these standards. To prevent any potential legal problems, the preemployment physical ability test must be task related and validated by a third party outside the fire department or government organization. The Equal Employment Opportunity Commission (EEOC) provides guidelines for establishing validated employment standards. The Candidate Physical Ability Test meets the requirements of the EEOC and has been approved by the fire service in a joint labor/management wellness/fitness initiative. The EEOC also allows validated standards to be transported from one jurisdiction to another. For instance, employment standards developed in one city can be applied in other jurisdictions having similar operational functions. The EEOC does require that specific preconditions be met to ensure that the employment standards are applicable when adopted from another jurisdiction. In addition to the EEOC and CPAT guidelines, the health and safety officer should be familiar with the impact of the Americans with Disabilities Act (ADA) and with state/provincial civil service regulations.

The preemployment physical ability test should consist of a representative number of evolutions that duplicate tasks from the task analysis (Figure 3.5). Average completion times for the purpose of scoring shall be determined by the average scores of current employees doing the same test. The following are examples of validated preemployment physical ability evolutions:

- **Step mill:** Simulates continuous stair climbing — an activity that firefighters may have to perform at multilevel incidents

- **Hose drag:** Simulates actions necessary to maneuver a fully charged fire hose

- **Victim search:** Simulates actions necessary to crawl and search a smoke-filled structure

Figure 3.5 The health and safety officer must ensure that new personnel are capable of performing the tasks outlined in the task analysis. *Courtesy of Mike Nixon.*

- **Victim rescue:** Simulates actions necessary to drag an unconscious, 130-pound (58.97 kg) victim to safety

- **Forcible entry:** Simulates forcing open a door to gain entry at an incident scene

- **Ladder:** Simulates various activities involved in the use of a regulation two-section ground ladder

- **Ceiling hook:** Simulates the use of a pike pole or ceiling hook in the removal of a ceiling or upper wall

Each evolution is timed, and the candidates are evaluated. The score is then used with the preemployment medical examination to determine the candidate's hiring status.

Annual Evaluations of Physical Fitness

NFPA 1500 requires the fire department to establish and provide a physical fitness program that meets the requirements of NFPA 1583, *Standard on Health-Related Fitness Programs for Fire Fighters*, 2000 edition, to enable members to develop and maintain an appropriate level of fitness to safely perform their assigned functions. The maintenance of fitness levels shall be based on fitness standards determined by the fire department physician. These levels must reflect the firefighter's assigned functions and activities and the severity of occupational injuries and illnesses associated with these activities. Results are compiled in a personnel file for each employee and maintained for analysis purposes.

All members of the department are required to participate in the physical fitness program (Figure 3.6). They shall be annually evaluated and certified to perform their assigned duties in emergency operations. Members who cannot meet the physical performance requirements shall enter physical rehabilitation programs to assist them in meeting their designated levels. The health and safety officer, fire department physician, or physical fitness officer shall devise a set of exercises and a schedule to assist the employee in meeting the goal.

It should be remembered that a physical fitness program is holistic, positive, rehabilitating, and educational. It is not punitive and should not be used in such a manner. The goal of physical fitness is to improve the quality of life for all firefighters and to help them live a long, healthy life. Good physical fitness will also improve the quality of service provided to the community.

Medical Evaluations

In addition to the preemployment physical ability test, a preemployment medical evaluation and full medical examination must be performed by the fire department physician or an appointed medical authority. This medical evaluation along with the examination determine whether the candidate meets a minimum level of health and wellness. The medical examination provides a baseline for all future medical evaluations and examinations (Figure 3.7). Minimum standards are outlined in NFPA 1582.

Following employment with the fire department, an annual medical evaluation shall be performed on each employee involved in emergency response operations. This evaluation can be in conjunction with the annual physical ability test or be separate. A medical evaluation is required following a lost-time injury or illness and the associated rehabilitation program. The employee must also receive a medical evaluation before returning to duty. Under the mandates of workers' compensation, strict procedures must be followed in the event of a member's occupational injury or illness and return to duty. A department should have procedures for reporting nonoccupational injuries and illnesses and the return to work.

The annual medical evaluation shall consist of the following items:

- Interval medical history

- Interval occupational history, including significant exposures

- Height and weight

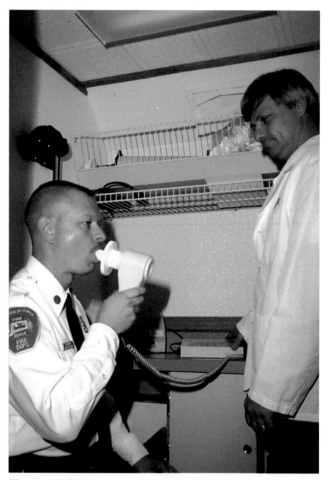

Figure 3.7 The pulmonary examination provides a baseline for all future respiratory examinations and is mandatory for all personnel using SCBA.

Figure 3.6 All personnel are required to undergo a full medical evaluation by the fire department physician.

- Blood pressure

- Heart rate and rhythm

In addition, a medical examination shall be required periodically depending on the age of the employee. The schedule required by NFPA 1582 is as follows:

- Age 29 and under — every 3 years

- Age 30 to 39 — every 2 years

- Age 40 and over — every year

Medical Examinations

The preemployment medical examination and the periodic medical examinations for current employees should include examination of the following components:

- Vital signs such as pulse, respiration, blood pressure, and temperature

- Dermatological system

- Ears, eyes, nose, throat, and mouth

- Cardiovascular system

- Respiratory system

- Gastrointestinal system

- Genitourinary system

- Endocrine and metabolic systems

- Musculoskeletal system

- Neurological system

If the need is indicated, the following tests may also be required:

- Audiometry testing

- Visual acuity and peripheral vision testing (Figure 3.8)

- Pulmonary function testing

- Laboratory testing

- Diagnostic imaging

- Electrocardiography

The list of Category A and Category B medical conditions is extensive and can be found in NFPA 1582. Category A medical conditions are defined as any medical condition that would keep a person

Figure 3.8 Visual acuity and peripheral vision testing may also be required as part of the annual medical examination.

from performing the duties of a firefighter in training or emergency operations. Conditions that may present a significant risk to the individual or other personnel include the following:

- Pulmonary hypertension

- Active tuberculosis

- Cerebral arteriosclerosis

- History of incapacitating hypoglycemia

- Angina pectoris

- Structural abnormality

- Limitations of motion in a joint

Candidates or current employees who have Category A medical conditions that would prevent them from performing the duties of a firefighter shall not be certified as fit for duty. Personnel with severe Category B medical conditions that would prevent them from performing the duties of a firefighter shall not be certified as fit for duty. However, the authority having jurisdiction can make reasonable accommodations to allow such an individual to perform other assigned functions. Some examples of Category B medical conditions include (but are not limited to) the following:

- Unequal hearing loss

- Allergic respiratory disorder

- Recurrent sinusitis

- Hernia

- Hypertension

- Diseases of the kidney, bladder, or prostate

- Use of steroids, stimulants, or narcotics

In the case of Category A and Category B medical problems, the fire department must take into consideration the legal implications of denying employment to a candidate or terminating an existing employee. Due to federal legislation, such as *Individuals with Handicaps or Disabilities*, Rehabilitation Act of 1973, as amended, Title 29 U. S. C. 791; Americans with Disabilities Act of 1990, Title 42 U. S. C. (29 *CFR* 1630.2 {n} {3}); Title VII of the Civil Rights Act of 1964, as amended, Title 42 U. S. C. 2000e, and numerous antidiscrimination laws, it is highly suggested that the advice of counsel be sought in cases of this nature.

Medical Records

To protect the employee, the Occupational Safety and Health Administration (OSHA) has very strict requirements regarding the development and maintenance of medical records as contained in Title 29 *CFR* 1910.20, *Medical Record Keeping*. In all cases, confidentiality is a key issue for the proper management of files and databases. The fire department physician and the employee are the only individuals who have access to an individual's file. The fire department physician provides the department with only an opinion concerning an individual's ability to perform the required duties and not a detailed description of any malady. Another important requirement is maintaining the personal medical files for the length of employment plus 30 years. The reason for this requirement is that if exposure to a hazardous or toxic chemical or a communicable disease occurs, all members involved who are experiencing an illness due to this particular incident can be tracked. If an employee with more than one year of service leaves the department for another position, a copy of the medical records should accompany the individual. Thorough record keeping is also necessary from a legal standpoint because it supports the position of the authority having jurisdiction in cases involving termination, duty status, and denial of employment to candidates.

Chapter 8, Section 4 of NFPA 1500 requires that the fire department maintain a confidential health file on each employee. Sample medical records forms are contained in Appendix E of NFPA 1582.

These health records need to incorporate details and results of the following events:

- Any and all medical evaluations and examinations
- Fitness evaluations
- Exposures (real or perceived) to hazardous materials and toxic chemicals
- Exposures (real or perceived) to blood, body fluids, or other potentially infectious materials
- Inhalations of airborne pathogens
- Excessive exposure to noise
- Occupational injuries or illnesses
- Other pertinent information that needs to be included in a member's health record

 Physical Fitness Programs

Before exercise can begin, an assessment must be made of the individual's level of fitness. This assessment allows the health and safety officer or health fitness coordinator to determine the correct exercise plan to meet the individual's needs. It also allows the officer or coordinator to establish a baseline for evaluating progress, to set a realistic range of expectations or goals, and to help the participants remain motivated.

The physical performance assessment is a series of exercises that the individual is required to perform. The individual's performance is then scored and compared to a predetermined scale. Once the level of fitness assessment is complete, a physical fitness program can be developed for each individual based on age, need, and gender. The United States Fire Administration (USFA) has developed the *Physical Fitness Coordinator's Manual for Fire Departments* (FA-95, 1990) that contains a detailed description of the assessment process and recommended exercises.

The physical fitness component of the health and wellness program must address flexibility, cardiovascular fitness, muscular fitness, and body composition. The health and safety officer will implement specific exercises using appropriate equipment to improve the individual's deficiencies in each of these areas.

Flexibility

Flexibility, or stretching, exercises are intended to improve the body's mobility (Figure 3.9). The flexibility portion is composed of five individual exercises to determine the range of motion of joints. These exercises include a sit-up and reach exercise, a back-of-thigh exercise, a shoulder exercise, a chest exercise, and a front-of-thigh exercise. Some recommended exercises include the following:

- Lower body stretches
  — Hamstring stretch
  — Quadriceps stretch
  — Toe tower or stretch
  — Lower back stretch
  — Abdominal stretch
  — Inner thigh stretch
  — Calf stretch
  — Crossover hip stretch
- Upper body stretches
  — Overhead shoulder and chest stretch
  — Neck stretch
  — Chest, back, and shoulder stretch
  — Biceps stretch
  — Triceps stretch
  — Crossover shoulder stretch
  — Back, waist, and shoulder stretch

The health and safety officer should break down the flexibility training program into attainable objectives that ultimately reach the goal established in the assessment. Attainable objectives may include a minimum number of sit-ups within a specified amount of time. These objectives are increased periodically, such as weekly, until the maintenance level goal is attained (Figure 3.10).

Cardiovascular Fitness

Due to the nature of fire fighting, flexibility is not the only requirement for firefighters. They also must have a high level of cardiovascular fitness, which determines the ability of the body to supply oxygen through the circulation system to the muscles while performing vigorous activities over an extended period of time. During the physical fitness assessment, the individual performs a 15-minute walk, jog, and/or run. The individual attempts to travel as far as possible during this time limit. A chart is used to determine the maximum oxygen consumption based on the distance completed. Once the assessment is complete, an exercise program that will maintain or improve the individual's ability is developed. To attain a level of fitness necessary to perform these activities, the cardiovascular exercises must be frequent (a minimum of three times a week); intense, at a moderate level of exertion; sustained for a minimum of 20 minutes; and of a type that exercises the large muscle groups continuously. Recommended exercises include the following:

- Jogging
- Speed walking
- Bicycling
- Stair climbing

Figure 3.9 Flexing or stretching exercises improve firefighter body mobility.

Figure 3.10 Weight machines can be provided for firefighters at the station or at a centralized location such as the training center.

- Skipping rope
- Swimming
- Rowing
- Cross-country skiing

Playing sports is not a recommended substitute for any of these exercises although it may be used to supplement them and make the training program more interesting (Figure 3.11).

Muscular Fitness

Good muscular fitness allows the firefighter to perform assigned tasks more effectively and efficiently and reduces the potential of personal injury. Adhering to a complete physical fitness program helps to reduce the potential for fatigue, heart disease, stroke, shortness of breath, and stress. To increase muscular fitness, the program needs to address strength, power, and endurance through the ability to perform push-ups, sit-ups, jump and reach exercises, and wall sits exercises. Definitions of these fitness terms are as follows:

- *Strength* is the maximum amount of force a muscle can generate.
- *Power* is the ability to exert strength quickly.
- *Endurance* is the ability to perform the activity repeatedly.

As the individual's strength increases through exercise, power and endurance also increase. Exercises are directed at the following six muscle groups (midsection, legs, arms, chest, shoulders, and back):

- Midsection
  — Curl-ups
  — Oblique curl-ups
- Legs (Figure 3.12)
  — Barbell or dumbbell half-squats
  — Barbell or dumbbell lunge
  — Barbell or dumbbell toe raise
  — Manual resistance knee extension
  — Manual resistance knee flex
- Arms (Figure 3.13)
  — Barbell or dumbbell biceps curl
  — Barbell or dumbbell triceps extension
  — Manual resistance biceps curl
  — Manual resistance triceps extension
- Chest
  — Barbell bench press

Figure 3.12 Leg presses are used to improve muscular fitness in the legs.

Figure 3.11 Competitive sports, such as basketball, can be used to supplement the physical fitness program.

Figure 3.13 The bench press helps to improve arm and chest muscles.

— Dumbbell flies

— Manual resistance push-ups

— Manual resistance flies

- Shoulders

  — Manual resistance lateral raise

  — Dumbbell lateral raise

- Back (Figure 3.14)

  — Manual resistance rowing

  — Dumbbell one-arm rowing

In all types of exercise, it is always important to warm up and cool down before and after each workout. These activities reduce the potential of injuries occurring during the exercise program (Figure 3.15).

Body Composition

Body composition, which is the relationship of lean-to-fat tissues in the body, is determined through comparing height to weight. The resulting number is converted to points that are added to the points achieved during the exercise portion of the assessment. The exercise portion of the physical fitness program helps to improve the muscle fitness of the individual. However, to reduce the quantity of fat tissues in the body, a healthy diet must also be utilized. The health and safety officer can influence diet through firefighter training in proper nutrition and eating habits.

Equipment

The equipment necessary for a physical fitness program varies depending on the finances available to

the department. At a minimum, it should consist of workout uniforms, T-shirts, jogging shorts, warm-up suits, and athletic shoes to be worn by participating members (Figure 3.16). In addition,

Figure 3.15 A proper warm up before each workout helps reduce the potential for injury.

Figure 3.14 One-arm rowing exercise improves upper back strength.

Figure 3.16 The department is responsible for providing personnel with proper workout clothing including shorts, shirts, warm-ups, and athletic shoes.

workout mats should be available in each station or at a central location where the exercises will take place. Free weights, exercise machines, powered stair-climbing equipment, and stationary bicycles may also be provided. The department might find it more economical to initially enter into a contract with a local health care facility or gymnasium for the use of its weights and machines. A physical fitness program can either cost the department a small amount of money or it can be a large expenditure depending on the approach taken. Either way, the cost is justified in the health and well-being of the members and the reduced cost of lost-time injuries.

Wellness Considerations

Going hand in hand with the physical fitness program, a good wellness program is necessary in the holistic approach. Wellness is a matter of education in the cause and effect of lifestyle on the body. Through a thorough nutrition, back care, and heart and lung diseases training program, employees will have an opportunity to see the benefits of making the changes in their lifestyles.

Nutrition

The importance of a good, balanced diet cannot be stressed enough. Recent studies have linked poor diet to heart disease, cancer, diabetes, high blood pressure, high cholesterol, and other diseases. As mentioned earlier, obesity is a result of improper nutrition and affects approximately 34 million Americans. As part of the health and wellness program, the health and safety officer should provide members of the department with training regarding the importance of good nutrition (Figure 3.17). This training can include learning the negative effect of certain foods such as those high in unsaturated fats, guidelines for a balanced diet, the results of good nutrition, and effective weight-control diets. Some departments are considering providing a recommended daily meal plan for each station via electronic mail (E-mail). An effective educational program on nutrition can have the added bonus of altering the off-duty lifestyle of the employees and their families. Information on nutritional issues can be obtained from the Centers for Disease Control, the state/provincial or local health department, or the American Heart Association.

Back Care

Back injuries are common for Americans between the ages of 25 and 50. These injuries are most common in people who do not adhere to a regular physical fitness program. Due to the nature of fire fighting, back injuries rank just below heart disease as a cause of firefighter injury. Therefore, it is important for the health and safety officer to provide a back care training program for employees (Figure 3.18). This training would include learning the following information:

- Techniques for proper lifting and carrying
- Types of equipment that require assistance for carrying
- Methods for dragging hoselines
- Exercises for strengthening the back muscles

Heart and Lung Diseases

As part of the overall cardiovascular system, heart and lung diseases can be addressed jointly through education. Training should include learning the following information:

- Importance of unimpaired lung capacity and unobstructed blood circulation
- Causes of heart and lung diseases
- Methods for reducing the potential for these diseases

The health and safety officer must implement a policy for the use of SCBA in all toxic atmospheres and in suspected or unknown atmospheres. All fire stations must provide a smoke-free environment. To assist firefighters in efforts to stop smoking, a smoking-cessation program should be developed and implemented by the department's safety and health committee. Finally, an effective physical fitness program must be implemented.

Infection Control

From the perspective of management, nothing has had a greater impact on the fire service and emergency medical services (EMS) than Title 29 *CFR* 1910.1030, *Bloodborne Pathogens,* and NFPA 1581, *Standard on Fire Department Infection Control Program*. This OSHA regulation and NFPA standard have changed the delivery of patient care provided

Figure 3.17 The wellness program is a balance of physical fitness, good nutrition, and training.

Figure 3.18 Instruction in proper lifting techniques is essential to preventing back injuries.

by emergency medical service providers. These documents require fire and emergency medical service organizations to perform a risk assessment of their operations from the standpoint of personnel safety and health. This risk assessment covers

Figure 3.19 Personnel involved in medical emergencies must be properly protected from potential exposure to body fluids. *Courtesy of Mike Wieder.*

both emergency and nonemergency duties and operations (Figure 3.19).

An exposure to a communicable disease can occur just as easily during the cleaning and decontamination of equipment at the fire station as it can during the delivery of patient care at the incident scene. A common philosophy that must spread throughout the fire and EMS community is that all patients must be treated as though they have a communicable disease. Infection control identifies such issues as the following:

- Personal protective clothing
- Mechanical resuscitation equipment
- Preexposure vaccinations for personnel
- Training and education in infection control
- Development of standard operating procedures
- Medical requirements

The consequences of an exposure to a variety of communicable diseases can be devastating for the infected member, his family, the fire department, and the governmental jurisdiction. Fire and EMS personnel must place their safety first and foremost in each situation.

The OSHA bloodborne pathogens regulation applies to all occupational exposures to blood or other potentially infectious materials. An *occupational exposure* is defined as reasonably anticipated skin, eye, mucous membrane, or parenteral contact with blood or other potentially infectious materials that may result from the performance of an employee's duties.

In order to develop, implement, and manage an effective infection control program, it is important that fire service administrative managers and the health and safety officer understand the regulations that govern infection control. The primary components of this OSHA regulation are as follows:

- Work practices
- Development of an exposure control plan for members at risk
- Training and education in infection control
- Engineering controls
- Personal protective equipment
- Housekeeping practices
- Hepatitis B vaccination (Figure 3.20)
- Postexposure evaluation and follow-up
- Medical record keeping

The key to ensuring compliance with the regulation is the development of the exposure control program and the training of members about the process for preventing and controlling infection exposure. The department shall designate an infection control officer as outlined in NFPA 1581 to develop the program and provide the training. Though all the components of the exposure control program previously listed are critical to the health and welfare of the members, a written plan provides the necessary guidelines to ensure compliance and to ensure that personnel understand the infection control process. This process for controlling potential exposures is accomplished by a thorough training and education program. It includes under-

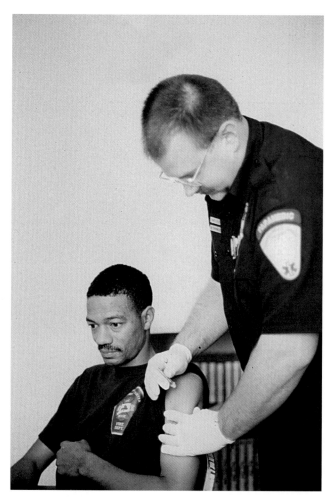

Figure 3.20 The department must provide members with inoculations against all strains of hepatitis.

standing potential exposure hazards, recognizing the appropriate level of protection to be worn or used, knowing the method for caring for an exposure victim, and reporting the exposure incident. Training shall occur when an individual is hired and then annually thereafter. Detailed contents for the training are found in Title 29 *CFR* 1910.1030.

Reporting

To meet the record-keeping requirements of Title 29 *CFR* 1910.1030, the fire department must maintain complete records of both the exposures to members and member training. The exposure report forms, completed at the time of the exposure, must contain the following:

- Name and Social Security number of the individual
- Copy of the individual's hepatitis B vaccination status

- Copy of the results of all examinations, medical testing, and follow-up procedures

- Copy of the fire department physician's written opinion

- Description of the employee's duties at the time of the incident

- Documentation of the circumstances of the exposure

- Results of the source individual's blood testing

- All medical records relevant to the treatment of the employee

NOTE: These records must be retained for 30 years from the date of termination of the employee. They must remain confidential and may only be released with the permission of the employee.

Training records for exposure awareness, which are retained for three years from the time of training, shall include the following:

- Dates of training sessions

- Summary of the training given

- Names and qualifications of the training personnel involved

- Names and job titles of all those attending training sessions

Follow-Up

The fire department physician, the infection control officer, and/or the health and safety officer shall ensure that medical evaluation and follow-up care occurs within 24 hours after the person's presumed exposure to bloodborne pathogens. The physician, the infection control officer, and/or the health and safety officer perform the following tasks during the evaluation:

- Document the route of exposure and the circumstances of exposure (infection control officer and/or health and safety officer).

- Identify the source individual if known (infection control officer and/or health and safety officer).

- Test the source individual's blood as soon as possible (physician or designated medical facility).

- Inform the exposed employee of the results of the testing of the source individual (physician, the infection control officer, and/or health and safety officer).

- Collect and test the exposed employee's blood for hepatitis viruses and HIV status (physician).

- Supply postexposure prophylaxis, if needed, to the employee (physician).

- Provide counseling to the employee (physician, infection control officer, and/or health and safety officer). The employee may be directed to the appropriate employee assistance program (EAP).

- Perform fitness-for-duty evaluation (physician).

Health and Safety Officer as Liaison

While the fire department physician is responsible for the medical aspects of the program, the health and safety officer provides the liaison between the administration and the fire department physician. The health and safety officer is also responsible for ensuring that exposure report information is collected by department officers or members and that the appropriate forms are completed.

One resource for infection control procedures is NFPA 1581. The NFPA 1581 requirements parallel the requirements of the OSHA bloodborne pathogens regulation but also provide specific criteria for an effective infection control program. Another source of information is the *Guide to Developing and Managing an Emergency Service Infection Control Program* by the USFA. This guide provides information on how to detail and implement a written infection control program. It also contains contacts for additional information.

 Hazardous Materials

Although incidents involving exposure to bloodborne pathogens are increasing, firefighters are regularly exposed to hazardous materials during emergency operations. The tasks involved with hazardous materials safety may be assigned either to the health and safety officer or to a hazardous materials branch safety officer. In either case, the skills necessary to perform these tasks are outlined in Chapter 8 of NFPA 472, *Standard for Professional Competence of Responders to Hazardous Materials Incidents.*

The health and safety officer must develop exposure protocols and implement them in the department's standard operating procedures manual. As with all other protocols, they must be reviewed and updated periodically and made a part of the department training program.

During a known hazardous materials incident, either the health and safety officer or the hazardous materials branch safety officer shall monitor the potential hazards to emergency personnel and advise the incident commander (Figure 3.21). Risks to consider during a hazardous materials incident, according to Title 29 *CFR* 1910.120, include the following:

- Exposures exceeding the permissible exposure limits and published exposure levels
- Immediately dangerous to life or health (IDLH) concentrations

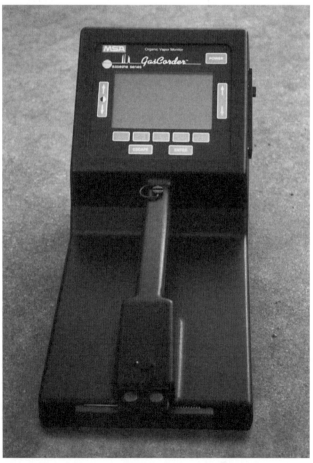

Figure 3.21 At hazardous materials incidents, the health and safety officer or the hazardous materials safety officer ensures that the air samples are monitored to prevent exposure to personnel. A toxicity meter similar to the one shown may be used.

- Potential skin absorption and irritation sources
- Potential eye irritation sources
- Oxygen deficiencies
- Explosion sensitivity and flammability ranges

Other duties of the health and safety officer during hazardous materials operations are found in Chapter 2, The Safety Officer as a Risk Manager, of this manual.

Should the presence of hazardous materials become known following an incident, Title 29 *CFR* 1910.120 requires that the fire department provide exposed employees with all available data on the materials to which they were exposed. Record keeping for hazardous materials exposure incidents should follow the same guidelines as those for bloodborne pathogens.

◆ Employee Issues

In the last half of the twentieth century, stress has become a common part of the lives of most people. Surveys indicate that one in five people experience stress daily. In the fire service, stress is a continuing fact of life from the moment the alarm sounds to the return to quarters. For that reason, a stress management program is an essential part of a health and wellness program. Stress in the fire service can take two forms: physiological stress and psychological stress.

Physiological Stress

Physiological stress is caused by the work environment and can be created by the following situations:

- Sound of the alert tone in quarters
- Shift work involving irregular schedules
- Lack of sleep or interrupted sleep cycles
- Loud noises (sirens, air horns, explosions, etc.)
- Station overcrowding
- Hazardous or toxic environments
- Exposure to extremes of heat and cold
- Long hours of physical exertion
- Carrying heavy personal protective equipment
- Fatigue (Figure 3.22)

Figure 3.22 Fatigue is a common physiological stress to firefighters. *Courtesy of Phoenix Fire Department.*

Environmental stress can be reduced by policies, education, and mitigation. Policies should be established that control the length of time a firefighter works during an incident and how often rehabilitation (rehab) must occur. The health and safety officer must mandate the use of SCBA during all phases of operations where toxic atmospheres are encountered. Personnel must be educated in recognizing their own limitations and not exceeding them.

The incident safety officer or health and safety officer on site must notify the incident commander when members or crews have reached their limit of physical effectiveness. In many cases, rest time is used as the only measurement as to when a person is able to return to work. Assessment of the person's baseline vital signs should be considered. Emergency medical technicians assigned to the rehabilitation unit gather this data and then provide it to the incident safety officer (Figure 3.23). Personnel are considered fit for duty when their body core temperatures and vital signs have returned to an acceptable level. The rehabilitation unit provides services such as nutrition, replacement fluids, and a rest area that allow personnel to return to acceptable physiological levels. Fire companies may not return to service until the incident safety officer releases them as fit for duty. In some cases, it may be necessary for the company to return to quarters for a change of uniform and dry protective clothing before returning to service. Firefighters have received burns from wearing damp or wet clothing under their protective clothing after working a succession of incidents.

Figure 3.23 The incident safety officer must ensure that all personnel receive rehabilitation before returning to duty. *Courtesy of Ron Jeffers.*

Psychological Stress

Psychological stress is more individual in nature and is created not only by the work environment but also by the total life experience of the firefighter. The following is a short list of psychological stressors that firefighters experience in their daily lives and take to work with them:

- Personality conflicts with coworkers
- Perceived lack of respect from managers, administration, or the public
- Boredom
- Lack of job satisfaction
- Concerns over promotion, layoffs, or retirement
- Feelings of inadequacy
- Fear of failure
- Personal injury to self or coworkers

- Death of coworker or family member
- Personal family problems

As part of the holistic health and wellness program, the health and safety officer must include activities that help members deal with stress. The National Fire Academy has developed a stress management model program that can be used to assist in the training of members. It defines stress, outlines the symptoms, lists the physical and psychological reactions, and provides activities for stress management. Four basic strategies for managing stress are as follows:

- Minimizing the stress-causing factors by avoiding, eliminating, or reducing them
- Changing how the individual perceives or views the problem
- Methods for relaxing both the mind and body
- Building the body's physical resistance to stress

The health and safety officer, in conjunction with the training division, should develop training programs that assist in meeting these four strategies. Recommended topics for training include the following:

- Time management
- Role management
- Conflict resolution
- Assertiveness training
- Relaxation methods
- Diet and nutrition guidelines

This training helps to prevent the possibility of psychological stress. However, should it occur, the health and safety officer must be able to recognize the symptoms and provide professional care through critical incident stress management, as outlined in the following section, or an employee assistance program, which is explained later in this chapter.

◆ Critical Incident Stress Management

Stress has always been a part of the firefighter's life due to the high level of uncertainty, limited control over the work environment, and the psychological

Figure 3.24 Firefighters experience emotional stress when exposed to incidents involving mass casualties and fatalities.

impact of repeated emergency calls. Add major events that exceed the normal level of stress and the ability of the body to cope, and critical stress develops (Figure 3.24). For this reason, a critical incident stress program must be part of the health and safety program of the department. This program is intended to manage the stress experienced by firefighters following incidents involving:

- Mass casualties
- Fatalities involving children
- Serious injury or fatalities involving members of the department
- Suicides
- Incidents involving close friends, relatives, or colleagues
- Violence directed toward firefighters
- Death of a civilian as a result of emergency operations
- Incidents generating excessive media attention

Signs and Symptoms

When the body undergoes normal levels of stress, it protects itself through increases in heart rate and blood pressure, oxygen consumption, muscle tension and strength, and the dilation of the pupils, among others. Excessive stress, however, results in further emotional and cognitive responses. Symptoms associated with excessive stress are as follows:

- Difficulty staying focused or concentrating
- Temporary loss of short-term memory

- Obsessive thoughts
- Loss of mental flexibility
- Tendency to withdraw or become isolated
- Feeling invulnerable
- Experiencing fantasies or wishful thinking
- Putting the mind on autopilot
- Abuse of alcohol and drugs

In order to address these incidents and the corresponding symptoms, the program should provide for a critical incident stress debriefing team (Figure 3.25). This team provides the necessary support for those members experiencing excessive stress. This team should consist of emergency services personnel, health care professionals, and clergy. The team members should all have specialized training in dealing with critical incident stress.

The critical incident stress debriefing team or an employee assistance program (EAP) provides training/support for emergency personnel in the causes and results of critical incident stress, stress reduction methods, and sources of professional assistance. At major incidents, team or program members provide guidance to the incident commander and the command staff and as well as support for individuals at the scene. Following a prolonged incident, team members help in the transition back to normal working conditions through the demobilization services. Debriefing the participants begins in this stage and continues into the defusing stage where team members discuss with the participants their reactions and feelings. The

Figure 3.25 Critical incident stress debriefing teams must be available whenever the need arises.

final phase, if necessary, is the formal debriefing meeting that occurs from one to three days following the event. Follow-up services, consisting of additional debriefing meetings, individual sessions, and referrals, can occur over a period of several months following the incident. Throughout the process, team members must maintain strict confidentiality.

Training

Although participation in the critical incident stress program is not mandatory, pre- and post-incident training can be used to prepare personnel for the possibility of a critical stress incident. The first step in reducing the effects of critical incident stress is to manage it by training firefighters about the psychological hazards of the job. This training should include information about critical incident stress, including its causes, symptoms, and some effective coping techniques. It is important to remember that this type of training is different from normal fire fighting training and is not exclusive to entry-level firefighters. All fire department and emergency services personnel should be included in critical incident stress management training.

◆ Postincident Illness and Injury

The potential for illness and injury at the emergency scene does not end with the completion of the operation. Members can develop symptoms of heat exhaustion, fatigue, stress, smoke inhalation, or other illnesses immediately following an emergency or upon returning to quarters. It is up to the incident safety officer or health and safety officer, if present, to recognize these symptoms and provide proper assistance.

Emergency Scene

At the incident scene, treatment can be provided through the rehab section where food, water, and cooldown materials are provided. In extreme cases, the health and safety officer or incident safety officer should recommend transportation of members to a medical facility for extended professional medical treatment. Transportation must be by ambulance and not by fire department apparatus or vehicles (Figure 3.26). Ambulances ensure constant medical attention is available during the trip.

Post-incident injury can occur during salvage and overhaul operations, when returning the companies to service, and during the return trip to quarters. The health and safety officer is responsible for mitigating potential injuries during these phases through education. Proper training in lifting, reaching, climbing, and carrying techniques reduces the potential for injuries. Driver education is essential for the safe movement of apparatus during the return trip to quarters. This education is attained through proper training and certification. If injuries do occur during this phase, the company officer and the health and safety officer, if present, are responsible for providing medical assistance and transportation of members to a health care facility. The health and safety officer must ensure

that all apparatus and department vehicles are equipped with an approved-sized medical first-aid box for minor injuries. The size of the kit is determined by the number of personnel assigned to the apparatus or vehicle at any given time, but it should not be smaller than one designed to serve four persons.

Station

Injuries and illness at the station can also be reduced through proper training and education. Personnel should be trained to apply the same work techniques, such as lifting, reaching, or climbing, in the station that they use at an incident scene. Education in proper hygiene, both personal and facility, should stress the importance of cleanliness to prevent the spread of diseases. The health and safety officer should develop a policy on food storage, dishwashing, clothes washing, and the use and cleaning of bedding. If possible, the fire department should provide bedding for each individual and have it changed following each shift.

The health and safety officer must ensure that decontamination sinks are provided for the cleaning and disinfecting of SCBA facepieces, medical equipment, and protective clothing (Figure 3.27).

Figure 3.26 Firefighters who require extended medical aid must be transported to a designated medical facility as soon as possible. *Courtesy of Ron Jeffers.*

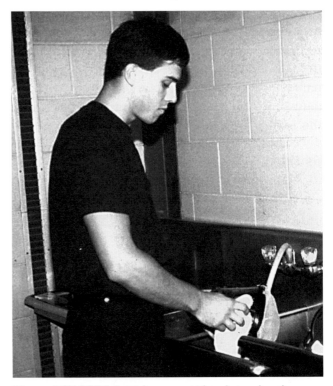

Figure 3.27 SCBA facepieces must be cleaned and disinfected in the designated area following each use.

Protective clothing (in particular turnout boots) must not be worn or taken into the living quarters. This requirement prevents the transmission of potentially hazardous waste into areas where personnel eat and sleep. Disinfectants should be provided for both personnel and equipment, and a policy written to control their use should be developed by either the health and safety committee or health and safety officer. Ventilated protective clothing storage must be provided in the apparatus bay or storage area. It is the responsibility of the fire department to provide decontamination sinks, disinfectants, and protective clothing storage areas in all fire department facilities.

The health and safety officer must ensure that the station is equipped with a first-aid kit designed for the number of occupants assigned to the facility at any given time. Additionally, an oxygen inhalator should be provided for each fire department facility. If the apparatus housed there is not normally equipped with an oxygen inhalator, the station should be so equipped.

 Employee Assistance

The levels of stress normally encountered by firefighters can result in dependence on tobacco products, abuse of alcohol or drugs, domestic violence, excessive gambling, and financial difficulties. In recognition of this fact, NFPA 1500, Chapter 9, mandates the establishment of an employee assistance program within the fire department. An employee assistance program is essential in the holistic approach to firefighter health and wellness. Because it is impossible for a fire department to provide the professional counseling services to meet these needs and ensure complete confidentiality, an EAP often includes outside contract counselors to provide the services. The health and safety officer is responsible for prevention, education, and referring employees to program counselors. He must also be aware of the symptoms of alcoholism, drug abuse, and other types of abuse in order to provide direction and care as soon as possible.

The effectiveness of the EAPs cannot be denied. Studies indicate that as many as 80 percent of those individuals who receive counseling return to full productive status within the workforce. EAPs can also reduce the cost of prolonged medical care and lost-time benefits. In addition to fire service members, the standard provides for assistance to immediate family members through the EAP. Finally, EAPs result in improved employee morale.

Substance Abuse

Of all the addictions of the American public, smoking and the use of tobacco products have received the greatest attention since the 1970s. It is estimated that over a quarter of the population over 25 years of age were smoking in 1990. This figure was a decrease from 33.5 percent in 1979. Smoking results in the loss of an estimated 400,000 lives in the United States every year. These deaths result from lung cancer, cardiovascular disease, and chronic obstructive lung cancer. According to the American Lung Association, smoking costs the nation approximately "$65 billion per year in health care and lost productivity." Surveys indicate that Canadians have experienced the same rates of disease due to smoking. Add the result of using smokeless tobacco products (mouth cancer, gum infections, tooth decay, and lowered sense of taste and smell), and the problem takes on greater magnitude.

Based on the national mortality rate due to smoking, the fire service loses approximately 1,800 members per year. Because firefighters are exposed to the unburned products of combustion resulting in chronic lung disease and loss of lung function, they are already at risk. Smoking, according to the American Lung Association, increases these hazards. Studies indicate that firefighters who smoke have a higher level of risk from heart and lung diseases than do firefighters who do not smoke. Smoking also results in lowered lung capacity and shortness of breath (Figure 3.28). These results can impair the stamina of firefighters during emergency operations such as wildland or high-rise incidents.

If smoking can have a negative effect on the health of the individual firefighter, then drug and alcohol abuse can have even a more widespread effect. Because drugs and alcohol impair judgment and slow reaction times, their impacts are not only on the individual but also on those around him, both uniformed and civilian. The United States Fire Administration estimates that as many as 10 percent of the 1.1 million firefighters in the United

Figure 3.28 Smoking can affect lung capacity and result in firefighter fatigue when using SCBA.

Figure 3.29 The health and safety officer is responsible for implementing a smoking-secession program including awareness classes.

States may be abusing drugs. The International Association of Fire Fighters estimates that 75 percent of the total firefighter population in the United States has used either drugs or alcohol. To offset the potential danger of this type of abuse, the fire department must establish a written policy within the EAP. The health and safety officer's duties in this area would involve developing and presenting education programs on alcohol and drug abuse, assisting the administration in developing a drug and alcohol policy, directing members to the EAP when necessary, and ensuring a liaison between the EAP and the administration.

Smoking Cessation

In order to create a healthy and smoke-free workplace, the fire department can approach the smoking problem from two directions. First, mandating that candidates be nonsmokers can be justified through existing documentation on the effects of smoking. Candidates can be required to be smoke free during their probationary employment. Many departments are implementing this preemployment criteria. Second, the department can provide smoking-cessation training for current members. These programs can be part of the training cycle for all members or specific cessation classes for individuals (Figure 3.29). The basis is an education program that points out the hazards of smoking, the reasons for quitting, and the methods available for quitting.

Counseling

The EAP outlined in department policy should have the ability to provide a wide range of counseling services for both the employee and his family. Besides smoking-cessation and substance-abuse assistance, domestic violence, child abuse, and family and financial counseling services should be available. Success of the program depends on the quality of the services offered, the support of the administration, the involvement of the members, and the confidentiality of all services. The health and safety officer should work to ensure that all of these elements are achieved.

Summary

An efficient and effective fire department depends on a healthy, physically fit, and emotionally stable membership. It is the health and safety officer's task to make physical fitness a reality through the implementation of a holistic health and wellness program. He must be able to address all areas of health and wellness both through prevention and mitigation (Figure 3.30). Through these efforts, the department can reduce injuries, reduce fatalities, reduce health costs, and improve the morale of the membership. The health and safety officer can also have a positive impact on the health, physical fitness, and longevity of the members of the department.

Figure 3.30 Fire department facilities should be smoke-free environments with well-ventilated designated smoking areas.

Health and Safety Officer's Responsibilities in the Training Process

4

This chapter provides information that will assist the reader in meeting the following job performance requirements from NFPA 1521, *Standard for Fire Department Safety Officer,* 1997 edition.

3-3.1 The health and safety officer shall ensure that training in safety procedures relating to all fire department operations and functions is provided to fire department members. Training shall address recommendations arising from the investigation of accidents, injuries, occupational deaths, illnesses, and exposures and the observation of incident scene activities.

3-3.2 The health and safety officer shall cause safety supervision to be provided for training activities, including all live burn exercises. All structural live burn exercises shall be conducted in accordance with NFPA 1403, *Standard on Live Fire Training Evolutions.* The health and safety officer or qualified designee shall be personally involved in preburn inspections of any acquired structures to be utilized for live fire training.

3-3.3 The health and safety officer shall develop and distribute safety and health information for the education of fire department members.

3-4.1 The health and safety officer shall manage an accident prevention program that addresses the items specified in this section. The health and safety officer shall be permitted to delegate the development, direct participation, review, or supervision of this program.

3-4.4 The health and safety officer shall periodically survey operations, procedures, equipment, and fire department facilities with regard to maintaining safe working practices and procedures. The health and safety officer shall report any recommendations to the fire chief or the fire chief's designated representative.

3-14.1 The health and safety officer shall develop procedures to ensure that safety and health issues are addressed during postincident analysis.

3-14.4 The health and safety officer shall include information about issues relating to the use of protective clothing and equipment, personnel accountability system, rehabilitation operations, and other issues affecting the safety and health of personnel at the incident scene.

Chapter 4

Health and Safety Officer's Responsibilities in the Training Process

For firefighters, safety begins with their initial training. Not only should they learn how to do their job safely and efficiently, but they should also be guaranteed that the training environment is safe. To this end, it is the responsibility of the health and safety officer to ensure that proper training in safety-related topics is provided, review the training procedures for unsafe conditions, and be present at potentially hazardous training events. The health and safety officer or incident safety officer should work closely with the administration, the training division, and the membership of the department to ensure that the training program meets the safety needs of the department as well as federal, state, and provincial mandates.

Role of the Health and Safety Officer in Training

NFPA 1521, Section 3.3, defines the role of the health and safety officer in three areas of training and education. First, the health and safety officer is responsible for ensuring that all members of the department receive training in safety procedures related to all operations and tasks of the department (Figure 4.1). Second, he is responsible for providing safety supervision of all training activities, in particular live burn exercises (Figure 4.2). Third, he is responsible for developing safety information for distribution to the members of the department. These responsibilities do not mean that the health and safety officer must personally participate in all training operations or teach all

Figure 4.1 The health and safety officer is responsible for ensuring that all members of the department receive training in safety procedures related to the operations of the department.

Figure 4.2 A safety officer shall be present at all training exercises that involve live fire exercises.

safety-related classes. Through delegation of authority to training officers, topic development by the health and safety committee, and "train-the-trainer" programs, the health and safety officer can create a systematic approach to training and education.

Interaction with the Training Division

Training is a systematic approach to maintaining the education, knowledge, skills, abilities, and competencies required for all members of the department. Training requirements must be outlined in a matrix that all members can understand and strive to complete. While the training division is responsible for the creation of a compliant training program, the health and safety officer is responsible for ensuring that the tasks and evolutions that are taught are in themselves safe. Training safety can be accomplished in a two-fold systematic approach.

First, the health and safety officer reviews all training evolutions in comparison to the task analysis developed to set hiring standards (see Chapter 3, Physical Fitness and Wellness Considerations). The training evolutions should conform to the operational tasks performed by the members of the department on a daily basis. The health and safety officer then ensures that the tasks are performed in as safe a manner as possible. For instance, one task defined in the task analysis that is common to the fire service is the raising or deploying of a two-section extension ladder by one or two firefighters (Figure 4.3). The health and safety officer reviews the method in which firefighters are trained to lift, carry, set, extend, and retract the ladder and recommends changes to the training programs if necessary. He takes into consideration the possibility of back injuries that can occur while lifting and extending, injuries that can occur during carrying, and potential environmental hazards associated with operating metal ladders. The health and safety officer performs such a review for each department standard, evolution, and practice.

Second, the health and safety officer provides the members of the training division with training in safety-related topics. Training officers should have the knowledge to function as incident safety officers during training operations. This knowledge relieves the health and safety officer of the need to be present during all training activities. The training officers should be able to recognize potential safety hazards, evaluate the need for proper infection control, understand and apply all vehicle operation laws, and follow the procedures outlined in the health and safety program. This training has an additional benefit of providing backup personnel for the health and safety officer. This procedure provides a core of trained personnel to function as incident safety officers at major events as well.

It should be recognized that all fire departments cannot afford to have a separate health and safety officer. When this situation occurs, the duties of the health and safety officer may be assigned to a training officer. This assignment is a logical use of existing skills because training and education are primary functions of both positions. It is important to delineate and separate the duties of health and safety officer and training officer when performed by the same person. Supervisors and commanding officers must recognize the importance of the separation of duties and allow the individual to fulfill the assigned task of the respective position. This is

Figure 4.3 The health and safety officer must be familiar with the training exercises to ensure that they are performed properly and safely.

especially important at live fire training exercises and emergency operations. The safety role takes precedence during these events. NFPA 1521 mandates that the health and safety officer not be assigned any conflicting tasks under these situations. It is also important to delineate the two functions during incident investigations and postincident analysis. As the health and safety officer, the individual must be able to perform an analysis free from outside pressures that might occur when there is a conflict of interest.

To properly and completely train emergency personnel for all potential hazards, recruit and in-service training must be as realistic as possible. Therefore, the approach to safety on the training ground mirrors the approach to safety at a real-life incident. Mandatory safety procedures on the training ground can develop into safe habits during the rest of a firefighter's career. Safety in training is essential to safety on the fireground. When safety is emphasized and learned in training, it tends to be remembered and practiced on the fireground.

Training Environment

The health and safety officer is responsible for surveying the general conditions of the training environment and recommending to the training division any changes that could reduce potential hazards (Figure 4.4). Such hazards include (but are not limited to) the following:

- Electric power lines (both overhead and underground)
- Damaged threads on hydrants, standpipes, and fire department connections
- Unsafe structures
- Limited means of egress from live burn structures
- Explosion hazards
- Asbestos in live burn structures
- Damaged training equipment such as ladders, rescue ropes, and nozzles
- Improperly cleaned equipment such as resuscitation dummies
- Noisy generators, fans, and power equipment
- Potential traffic hazards around remote training sites

In addition, the health and safety officer should review each training exercise for its impact on the surrounding environment. Environmental issues include the following:

- Noise pollution
- Ground water pollution by runoff of hazardous waste
- Air pollution from smoke and other contaminants
- Soil pollution
- Psychological or emotional effect on neighbors

The health and safety officer must also consider the impact of the environment on the personnel involved in the training. The argument is often made that firefighters must be prepared to engage in emergency operations in all types of weather. This may be true; however, training in harsh weather must not endanger the firefighter's health and safety. The health and safety officer advises the training division when, in his opinion, the weather conditions would have an unsafe effect on training. High winds, lightning, sleet, heavy fog, or icy roads are obvious conditions that would warrant the canceling of training operations. Conditions that would make it difficult to drive to the training site should also be considered. These considerations include days when the weather conditions could cause an increase in ozone emissions that would damage the environment. The health and safety officer also considers factors such as the windchill index and the heat stress index. By teaching the members of

Figure 4.4 Flashover simulators and live fire simulators must be inspected prior to use to ensure that they are in proper working condition.

the training division to be aware of environmental hazards, the health and safety officer can delegate these decisions to them.

During nonfire-training exercises, the health and safety officer must ensure that proper protection is worn and that protocol is used. Eye, hand, and hearing protection are mandatory when operating power tools such as rotary saws, power extrication equipment, and chain saws. The health and safety officer must be aware of potential sparking hazards in the presence of flammable liquids and gases. NFPA 1500 stresses the importance of safety in every aspect of tool and equipment design, construction, purchase, usage, maintenance, inspection, and repair (Figure 4.5). It is up to the health and safety officer to ensure that this standard is complied with both on the training ground and at emergency incidents.

Live Burn Exercises

To make training as realistic as possible, many fire departments use live burn exercises to expose their new recruits and current members to structural and nonstructural fires. NFPA 1521 requires that the health and safety officer perform the functions for live burns outlined in NFPA 1403, *Standard on Live Fire Training Evolutions*. According to this standard on live burns, "a safety officer shall be appointed for all live fire training evolutions." Live burn training exercises include the following components:

* Structures
  — Specifically designed burn buildings
  — Acquired structures located at a remote site
* Gas-fired permanent training structures (Figure 4.6)
* Nongas-fired permanent training structures
* Exterior props
  — Flashover simulators
  — Vehicles
  — Trailers
  — Railroad tank cars
  — Class A combustibles
* Exterior Class B fires
  — Liquefied petroleum gas
  — Flammable/combustible liquids

Figure 4.5 Training equipment, such as positive pressure ventilation fans, must be inspected to ensure safe operation.

Figure 4.6 Gas-fired simulators require safety inspections prior to use.

Regardless of the type of live burn, the primary duties of the health and safety officer at a live burn are the following:

* To prevent unsafe acts from occurring
* To eliminate any unsafe conditions

Because the health and safety officer is responsible for the safety of all those present, including students, instructors, visitors, and spectators, he shall not be assigned any other duties. If the live burn is large or complex, additional safety personnel should be assigned. This procedure provides greater control and creates the same type of safety management found in IMS-type operations. Incident safety officers should also be present at live burn incidents. In addition, the health and safety officer has the authority to "intervene and control" any part of the operation that he believes to be unsafe or life threatening. This authority is identical to that given him and the incident safety officer in NFPA 1521 at emergency scenes.

Although the primary duties of the health and safety officer remain the same for all live burns, each type of live burn has its own requirements for the health and safety officer. Working in conjunction with the training division, the health and safety officer must make certain that a burn structure meets a specified level of compliance with the NFPA 1403 standard. This compliance is especially critical with structures acquired from outside owners and located away from the training facility (Figure 4.7). The health and safety officer is responsible for the following tasks to ensure that acquired structures are in compliance with NFPA 1403:

- Ensure the structural integrity of the building.

- Remove hazardous materials from the structure.

- Repair structural members that may create a hazard.
  — Repair stair treads, risers, and railings.
  — Secure holes in floors.
  — Secure loose floorboards.
  — Secure or patch walls and ceilings.
  — Secure or remove loose bricks in masonry walls or chimneys.
  — Provide adequate roof level ventilation.

- Shut off utilities.

- Remove trash and debris that may cause a hazard.

- Remove low-density combustible fiberboard wall coverings.

- Remove vermin, insects, and toxic vegetation from the structure and surrounding area.

- Provide exposure protection for adjacent structures.

- Remove vegetation from the area of the burn building.

- Provide sufficient egress routes.

- Develop a predetermined evacuation plan.

- Ensure that all participants understand the incident management system to be used.

Any building or structure that cannot be made safe may not be used for interior structural fire-fighting training.

Other factors that the health and safety officer must consider are vehicular and pedestrian traffic in the area, sufficient water supply, and weather conditions during the exercise. The exercise should be approached in the same manner as a real incident with the health and safety officer performing the same duties as those assigned at an emergency. This procedure provides him with training in establishing a rehabilitation unit and control zones and working with the incident commander during the training.

The health and safety officer monitors the operation and advises the incident commander on potential safety hazards or unsafe acts. He monitors the backup hoselines and rapid intervention crews; the wearing of proper personal protective clothing, SCBA, and personal alarm devices; the accountability system; and the progress and development of the fire (Figure 4.8). He should be particularly aware of the physiological condition of

Figure 4.7 Vacant buildings that are donated to the fire department for training fires must be inspected and hazards eliminated before training exercises.

Figure 4.8 Live fire exercises present a training opportunity for incident safety officers as well as firefighters.

the participants including exhaustion, heat stress, fatigue, and physical condition. He is responsible for establishing a rehabilitation section including standby EMS personnel.

Where Class B flammable liquids and liquefied petroleum gas fires are used for training, the health and safety officer determines that the minimum quantity of flammable material is used to provide the desired quantity of fire. He must also see that the correct extinguishing agent is on hand and that the quantity is sufficient for both the training and for backup. All personnel must wear complete protective clothing with SCBA and personal alarm devices during the training exercise.

While it is the responsibility of the training division to maintain records on the live burn exercise, the health and safety officer prepares the same reports on the safety aspect of the training as he does at an emergency incident. The health and safety officer will be asked to participate in a post-training critique and must be able to document the need for any changes based on safety considerations. This information provides an indication of the effectiveness of the safety-training program and highlights any trends in unsafe acts. These records of training incidents are maintained along with the emergency incident records kept by the health and safety officer.

Interaction with the Company Officer

Safety training at the company level can be accomplished in two ways. First, the health and safety officer can provide periodic training at the fire station (Figure 4.9). This training can be either the result of mandated job-wide training on a specific topic such as infection control or the result of an unsafe act such as improper backing of an apparatus on the part of a single company. In either case, the health and safety officer works directly with the company officer to provide the required training material.

A second method of providing company-level safety training is the "train-the-trainer" approach. In this case, the health and safety officer trains all company officers, providing them with outlines, visual aids, and testing material, and has them lead the training for their individual companies. This

type of training might lend itself to changes in department safety rules and regulations or operational procedures.

In smaller departments, the health and safety officer may also be a company officer. When this situation occurs, the job requirements must be well defined. The person fulfilling these dual duties must be well trained in all aspects of safety, health and wellness, infection control, and hazardous materials. Without proper education and administrative support, the company-level health and safety officer will be unprepared to accomplish the duties of this position.

The Health and Safety Officer and the Training Process

To effectively communicate safety information, the health and safety officer must have knowledge specific to the training function. This knowledge includes curriculum development, teaching theory, communication skills, and a working knowledge of the equipment/training aids used in training.

Figure 4.9 Infection control and other safety-related training may take place within the fire station. *Courtesy of Mike Nixon.*

Curriculum Development

Curriculum development involves the logical structuring of the course content. The health and safety officer must be able to prepare a course outline that includes the necessary facts, data, and information but does not overload the student with unnecessary material. At the same time, the material must be complete and understandable. Curriculum development requires that the health and safety officer determine the desired outcome or results and know how to judge them before preparing the outline. The health and safety officer must know the audience, define the topic of the instruction, and gather supporting evidence. The next step of curriculum development is to prepare an outline and produce the necessary visual aids. Because this is the same curriculum development process that the training division uses, the health and safety officer should use training officers as a resource. The health and safety officer may also choose to contact the local public school system for guidance, attend a college course in education or teaching theory, or attend an educational methodology course offered by his state/provincial fire training academy or the National Fire Academy. The IFSTA **Instructor** manual is also a rich source for teaching and curriculum development information.

Teaching Theory

With the curriculum developed, the health and safety officer must decide on the best way to present the material. One format that can be implemented is the one used by the United States military and involves building in redundancy:

1. *Preview the material.* "Here is what I am going to tell you."

2. *Give the actual subject material.* "Here is the material."

3. *Review the material.* "Here is what I just told you."

This type of delivery can be effective for brief presentations on nontechnical subjects such as the importance of wearing protective clothing during overhaul. More complicated topics involving statistics or step-by-step processes do not necessarily lend themselves to this presentation style.

A more effective method is the one taught in public speaking courses for delivering an informative speech. This style is used when the topics are one of the following:

- Principles, theories, or abstract ideas such as the effects of toxins on the bloodstream

- Objects or tangible items such as the introduction of a new type of tool

- Procedures or process such as the proper use of power extrication equipment

- Events or actual occurrences such as a postincident critique

The organizational format in the informative style involves following these four steps:

Step 1: Set the stage, and gain the attention of the audience.

Step 2: Describe the situation or topic.

Step 3: Build to a logical climax.

Step 4: Bring the story to a conclusion, wrapping up all of the loose ends.

Another style follows the pattern of the persuasive speech. The goal of the instructor is to influence or change the following characteristics:

- Attitudes

- Beliefs

- Values

- Behavior

To accomplish this goal of persuasive speech, the instructor uses the following five-step organizational theory developed many years ago by Alan Monroe. The Monroe motivated sequence is a simple, yet effective strategy (Figure 4.10).

Step 1: *Attention*: Give a startling statement, personal experience, an unusual statistic, or an analogy that gets the attention of the audience.

Step 2: *Need*: Tell the audience what the problem is and why it needs to be solved. Include all evidence and data to support the need for a solution.

Step 3: *Satisfaction*: Describe how the solution solves the problem.

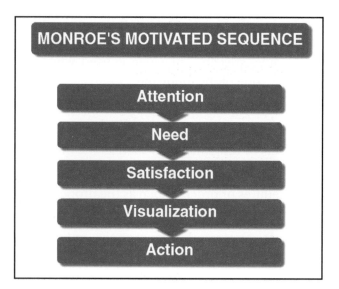

Figure 4.10 Monroe's motivated sequence provides the instructor with a model for efficiently and effectively teaching firefighters.

Figure 4.11 The effectiveness of the training process can be determined through the testing process.

Step 4: *Visualization*: Describe the results that occur if changes are made and if they are not made. Visualization allows the audience to see how a change, or lack of it, affects them.

Step 5: *Action*: Provide the audience with the steps they should take to accomplish the goal. Action is the conclusion that motivates the audience to make the change.

An example of a persuasive topic might be the importance of the two-in/two-out rule in confined-space incidents and interior structural fire fighting attacks. The instructor could begin the presentation by relating a personal experience involving a fire company trapped within a structure by flashover (***attention step***). He would then continue with a detailed description of the number of firefighters injured or killed while making interior attacks (***need step***). Next, the instructor explains the concept of the two-in/two-out rule and explains how the presence of a rapid intervention team can solve the problem (***satisfaction step***). Now, the audience is asked to describe the results of the change such as fewer injuries or deaths during confined space or interior fire fighting (***visualization step***). Finally, the audience is given the steps necessary to make the change such as a written policy requiring two-in/two-out or an increase in the number of personnel assigned to engine companies (***action step***). Other topics for the persuasive approach would include smoking cessation, wearing SCBA

during salvage and overhaul, understanding the advantages of good nutrition, or using an infection control procedure.

Once the material has been presented, the results must be judged for effectiveness. This judging is accomplished through a testing process (Figure 4.11). This process requires the health and safety officer to develop questions or skill evolutions that indicate how well the student understood and retained the material. Testing takes place at the end of the course and at periodic intervals of one or two years. If the test results indicate that a student did not fully comprehend the material, the health and safety officer should select a different method to present the material until the student successfully completes the testing process.

Training Equipment/Training Aids

Effective curriculum development and presentation skills are not the only things that the health and safety officer needs in order to be a good instructor. The health and safety officer must have a working knowledge of the various equipment and aids used in the training process in order to operate them properly and to evaluate their safety hazards in the training environment (Figure 4.12).

The health and safety officer, like the training officer, must be familiar with the various classroom-training aids available, including the following:

- TV/VCR units

- DVD projection equipment

Figure 4.12 The health and safety officer must be familiar with the audiovisual equipment used in the training process, including slide projectors and video equipment.

- Overhead projectors
- Slide projectors
- Motion picture projectors
- CD-ROM projection equipment
- Audio equipment
- Computer-equipped classrooms
- Closed-circuit television

Other training devices may include the following:
- Anatomical models
- Resuscitation mannequins
- Defibrillators
- Mock-ups of pumps, engines, etc.

Outside the classroom, the health and safety officer must be familiar with the operation of equipment used on the training ground. This equipment would include but is not limited to the following:
- Pump test stands

- Physical ability testing equipment
- Fire fighting tools and equipment
- Power tools and equipment
- Fire apparatus
- Hose, nozzles, and appliances
- Training center water supply system
- Gas-fired training structures

Familiarity with the equipment used in training allows the health and safety officer to provide instructions quickly and effectively. It also allows him to recognize potential safety hazards created by the equipment or the improper use of the equipment. Familiarity extends to the types of personal protective equipment and its use in training. Not all training is performed in station wear or full protective clothing. Some training such as high-angle rescue or confined-space entry requires specialized protection. The health and safety officer must be able to evaluate the proper type of hard hat, footwear, or coverall to be used in the training evolution. The type of personal protective clothing worn during emergency operations should also be worn during training.

The training process is as complex and involved as the safety process. Therefore, the health and safety officer must work closely with the training division in the development and delivery of safety-related training. At the same time, the health and safety officer is responsible for ensuring that the training process as a whole is safe for participants.

 Role of the Health and Safety Officer in Updating Training

The health and safety officer must take a proactive approach to correcting unsafe acts. He must be positive and present the need for change without placing blame for the unsafe act. Punishment is not part of his job description.

In order to stay proactive in the safety process, the health and safety officer is responsible for updating safety-related training and correcting potential safety hazards. This is done as a result of incident investigations, postincident analyses, changes in the trends in safety, and the health and safety committee input.

NFPA 1521 requires that the health and safety officer investigate all incidents that result in injury, hazardous exposure, death, illness, or other hazardous conditions. From the information compiled during this investigation, he analyzes the incident and recommends any necessary changes in operational procedures to the chief of the department. Once changes have been approved, they must be communicated to the members of the department through training or education.

Incident Investigation

The objective of the incident investigation is to determine the factors that contributed to firefighter fatality or injury or damage to departmental property. Investigations are fact-finding rather than faultfinding procedures. The health and safety officer must determine whether an unsafe act or negligence was a part of the incident (Figure 4.13). The purposes of this type of investigation include the following:

- Avoid loss of human resources and equipment.

- Ensure better cost-effectiveness in the use of personnel and equipment.

- Improve the morale of both the department's personnel and members of the community.

- Determine the change or deviation that caused the incident.

- Determine hazardous conditions to which fire and emergency services personnel may be exposed.

Figure 4.13 The health and safety officer is responsible for investigating all property loss incidents involving department vehicles. *Courtesy of Ron Jeffers.*

- Direct the attention of management or officers to the causes of incidents.

- Examine facts as though they have a legal impact on incident cases.

During the investigation phase of the incident, the health and safety officer collects information from many sources such as the following:

- Dispatch records

- Interviews with participants

- Emergency incident reports

- Incident action plan

- Incident safety plan

- Police reports

- Photographs and videotapes of the scene or incident

- Site diagrams

- Physical property involved in the incident

- Laboratory analysis of evidence relating to the incident

- Testimony from experts

- Incident safety officer report

This information provides the health and safety officer with a fairly accurate description of the incident. A written report is then compiled on the incident relating to health and safety issues. This report is the basis for the postincident analysis and any corrective actions that need to be taken by the department.

Postincident Analysis

The postincident analysis lists the cause of the incident, the department's actions in bringing the incident under control, and the effects of the incident. The health and safety officer is concerned with the health and safety aspects of the incident. He is concerned not only with the cause of firefighter fatalities and injuries but also with the proper use of protective clothing and respiratory protection equipment, establishment of rapid intervention crews, the use of PASS devices, the use of personnel accountability systems, and the establishment of rehabilitation services.

Case Studies

An example of safety issues that arose from an incident investigation occurred in a Midwestern city a number of years ago (Figure 4.14). Two apparatus (an engine and an aerial device) collided in an intersection while responding to the same incident. The aerial device struck the engine in the right side pump section causing the engine to turn over and come to rest upright, facing the opposite direction. Injuries to the two crews were minor, the worst being a dislocated shoulder. The investigation determined that the primary cause of the incident was operator error on the part of both operators. The engine operator failed to yield to a red traffic signal, and the aerial operator had not determined whether the intersection was clear before proceeding through it. A lack of communication was also apparent as neither company attempted to talk to the other about their routes during the response. Loose tools and equipment within the crew compartment of the engine had also created a potential hazard to the crew.

During the postincident analysis, the health and safety officer determined that the department's standard operating procedures had been violated (Figure 4.15). In addition, he determined that the unofficial practice of companies from the same station using different travel routes had contributed to the incident. The health and safety officer made recommendations for changes to the SOPs. Training was developed from the final report to ensure that a similar situation did not reoccur. As a result, the department had the two operators teach other personnel through training meetings and the use of closed-circuit television. In addition, the department implemented a program of relocating and securing all loose tools and equipment within the personnel compartment on all apparatus.

Figure 4.14 The health and safety officer is responsible for analyzing the cause of collisions involving department vehicles and recommending changes in response and operating procedures. *Courtesy of Mike Mallory.*

In another incident, changes that resulted from a postincident analysis many years ago revolved around personal protective clothing and an engine company responding before the crew was completely ready. The engine company responded to a second structure fire immediately after returning to service from an earlier working fire. During the second incident, one of the firefighters was trapped in what appeared to be a flashover event and received severe burns to his upper torso and arms.

The postincident analysis found that his protective clothing liner and station wear had been soaked with perspiration resulting in steam burns, especially in areas where the coat was compressed by SCBA straps. Changes that resulted from this analysis included a change in the design of the protective clothing to an improved liner system and reemphasizing the importance of having all members of the company prepared to respond to another emergency even if it involved returning to quarters for dry clothing (Figure 4.16).

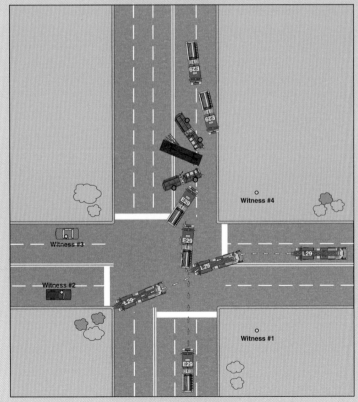

Figure 4.15 This diagram illustrates an incident involving two apparatus responding to the same location using converging routes. Lack of communications and failure to yield the right-of-way were contributing factors in the collision. *Based on a site sketch by Mike Mallory.*

Figure 4.16 Firefighters must be completely rehabilitated before they are allowed to return to duty. *Courtesy of Ron Jeffers.*

Staying Current

To stay current with trends in the safety field, the health and safety officer should obtain as much up-to-date information as possible. This updating can be accomplished through reading fire, health, and safety trade journals; attending safety conferences; being an active member in safety organizations; maintaining and analyzing incident safety data, and maintaining a positive relationship with equipment manufacturers. Other sources for safety information are the Learning Resource Center at the National Fire Academy and safety-related web sites on the Internet. Sites are operated by the NFPA, IFSTA, Centers for Disease Control, Department of Labor, OSHA, the International Association of Fire Fighters, the International Association of Fire Chiefs, and the National Volunteer Fire Council, among others. All sites contain fairly current information and documentation.

Health and Safety Committee

Finally, the safety-training program can benefit from the work of the health and safety committee. This committee can be a clearinghouse for safety-related issues that have not caused an incident (injury, illness, fatality, or property damage) but have the potential for an unsafe act. It also allows the membership of the department to express their concerns in an open atmosphere. This is an opportunity for the health and safety officer to maintain a good relationship with all members of the department. Training division representation is important if the health and safety committee is to be an effective influence on safety.

 ## Summary

For the fire department to remain a proactive force in the safety process, it must be able to eliminate unsafe acts and hazardous conditions as efficiently as possible. The health and safety officer, through the requirements of NFPA 1521, can affect this process. He must work with all members of the department to identify unsafe conditions, alter the habits that may have caused them, and educate the membership in corrective actions. By working closely with the training division, the health and safety officer can use a systematic team approach for a positive outcome.

5

Facilities Safety

Job Performance Requirements

This chapter provides information that will assist the reader in meeting the following job performance requirements from NFPA 1521, *Standard for Fire Department Safety Officer,* 1997 edition.

3-8.1 The health and safety officer shall ensure all fire department facilities are inspected in accordance with Chapter 7 of NFPA 1500, *Standard on Fire Department Occupational Safety and Health Program.*

3-8.2 The health and safety officer shall ensure that any safety or health hazards or code violations are corrected in a prompt and timely manner.

Chapter 5

Facilities Safety

The health and safety officer is part of the administrative team that is responsible for the creation and maintenance of a safe working environment for department personnel whether they are at an incident or in quarters. When dealing with safety issues in and around fire department facilities, the health and safety officer must ensure that the physical structure provides a safe environment. Safe environments are accomplished through the design and construction of new facilities (Figure 5.1) that meet all applicable codes and standards and through the inspection and renovation of older existing structures (Figure 5.2). The health and safety officer must also educate the members of the department in the safe use of their facilities, which includes proper cleaning, storage, and maintenance techniques. Because the fire department enforces similar safety-related fire codes within the community through building inspection and fire prevention inspection programs, it is important that all fire department facilities meet or exceed those same requirements (Figure 5.3).

Besides the safety requirements found in NFPA 1500 and 1521, NFPA 1581, *Standard on Fire Department Infection Control Program*, mandates safety and health-related policy, protocol, and design criteria for fire department facilities. Included in this standard are the requirements for an infection control officer, a written risk management plan for infection control, and training and education of all personnel in infection control. The infection control officer may be the health and safety officer or another officer who is delegated the responsibility for infection control. The written risk management plan includes identification, evaluation, control, and monitoring of risks associated with infection exposure in the following areas:

• While in fire department facilities

Figure 5.1 Modern fire stations are designed to provide a safe, healthy, and comfortable environment for employees.

Figure 5.2 Training facilities include classrooms, burn buildings, and areas for simulated tactical evolutions.

Figure 5.3 Many fire departments operate their own apparatus and vehicle maintenance shops, which create safety hazards unique to this type of use.

- On and around fire department apparatus

- While cleaning or disinfecting personal protective equipment, station wear uniforms, or emergency medical equipment

- Any other situations that might lead to exposure or contamination from infectious materials

The magnitude of health and safety problems associated with fire department facilities can be found in data compiled by the NFPA. Between 1983 and 1995, 17 firefighters died in fire stations as a result of causes other than cardiovascular systems deaths. Causes included falls (7), crushing traumas (3), electrocution (1), boiler explosions (1), carbon monoxide poisoning (2), SCBA cylinder explosions (1), and homicide (1). The number of facility-related injuries and illnesses are more difficult to determine due to the inconsistency of national data collection. Estimates can be based on information obtained from a cross section of individual departments (Table 5.1). The health and safety officer may want

to contact departments of a similar size to compare facility-related injuries and illnesses (Figure 5.4).

This chapter is intended to provide the health and safety guidelines for designing new facilities, renovating existing facilities, inspecting facilities, recognizing potential hazards in the facilities, and removing those hazards before they contribute to personnel injury or illness.

◆ Facility Design for Safety

As existing facilities grow older and the coverage area of a fire department increases, it becomes necessary for new facilities to be constructed. New structures may house administrative offices, training functions, supply and maintenance operations, or active fire companies. Whatever the use, the new facility must provide a safe working environment for the personnel housed there. It is critical that safety issues be addressed during the planning phase of the process.

Table 5.1
Rank of Fire Department Activities Resulting in Injuries — 1992-1994

Rank	Activity	Injuries by Year			
		1992	1993	1994	Total Injuries
1	Vehicle maintenance	42	37	30	109
2	Physical fitness activity	40	30	24	94
3	Moving about station, normal activity	23	29	19	71
4	Moving about station, alarm sounding	23	18	25	66
5	Station maintenance	19	12	17	48
6	Boarding apparatus	15	11	12	38
7	Cooking and food preparation	6	10	7	23
8	Sleeping (getting out of quarters)	8	1	8	17
9	Equipment maintenance	1	2	11	14
10	Other activity	0	3	6	9
11	Administrative work	2	4	2	8
12	Exiting apparatus	not listed	not listed	3	3
12	Physical fitness testing	3	0	0	3
12	Showering/personal hygiene	0	1	2	3
13	Training activity or drill	1	0	0	1
	Total	183	157	166	506

Data compiled by a Southwestern fire department with approximately 600 emergency response personnel.

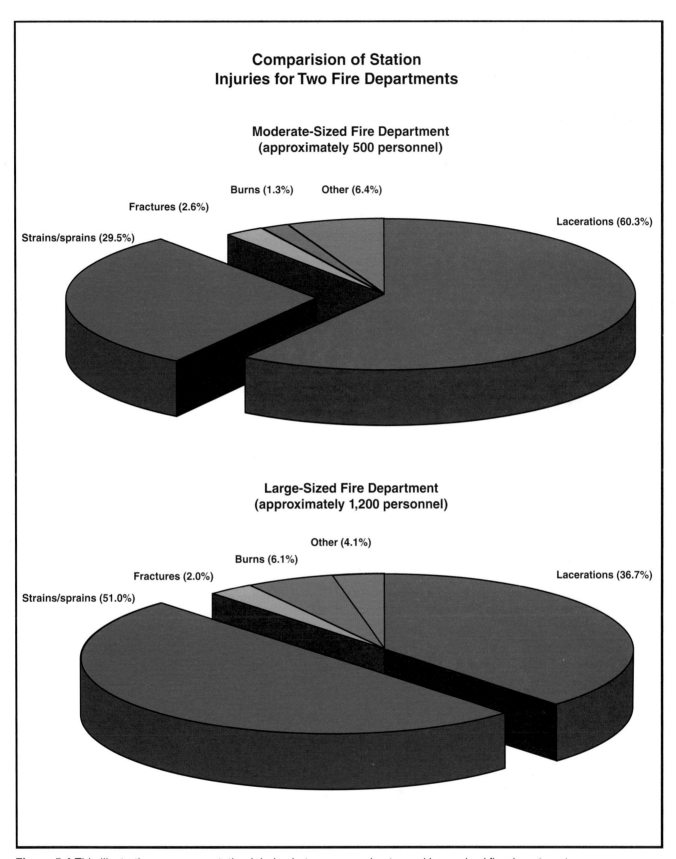

**Comparision of Station
Injuries for Two Fire Departments**

**Moderate-Sized Fire Department
(approximately 500 personnel)**

Burns (1.3%)   Other (6.4%)

Fractures (2.6%)

Strains/sprains (29.5%)

Lacerations (60.3%)

**Large-Sized Fire Department
(approximately 1,200 personnel)**

Other (4.1%)

Burns (6.1%)

Fractures (2.0%)

Strains/sprains (51.0%)

Lacerations (36.7%)

Figure 5.4 This illustration compares station injuries between a moderate- and large-sized fire department.

New Facility Design

When fire department officials determine that a new facility is needed, the health and safety officer should be part of the design team (Figure 5.5). His involvement in the planning and preconstruction process can ensure that safety hazards are eliminated before they occur. Many sources such as the health and safety committee; task analyses; inspections; applicable laws, codes, and standards; and security consideration guidelines are available to guide the health and safety officer in helping to develop a safe and efficient facility design.

Health and Safety Committee

One of the primary sources for design recommendations is the health and safety committee. This group represents the members of the department who know firsthand the limitations of the existing facilities and who will have to live with the new design. Brainstorming sessions with this group provides commonsense recommendations in addition to those found in the building codes or architectural design standards. Several issues, such as the best type of floor covering to use in the living quarters and the type of cleaning equipment required, may be discussed. Questions about the efficiency of storage, air circulation, or apparatus room floor drainage influences the final design and may come to light from this committee.

Figure 5.5 The health and safety officer should be a member of the department's facilities design committee.

Additionally, the authority having jurisdiction may wish to form a separate building or facilities committee. Representation on the committee should include at least one person from each of the following groups:

- *Department administration or staff personnel* — Provide a historical overview of what has worked and what hasn't worked for the department. They are also keenly aware of how the station must function to meet overall department operational needs.

- *Operational or line personnel who must live and work in these facilities* — Should have an active voice in the design process. These individuals provide the most beneficial information and know what it takes to make the structure work both functionally and comfortably.

- *Member organization representatives* — Speak for the line personnel and try to obtain as many features as possible for increasing their safety and comfort.

- *Budget officials* — Want the most economical structure for meeting the budget. They may not have projected funds to cover features based on requirements from standards that are voluntary. Failure to comply with existing federal, state/provincial, or local requirements during the initial planning process usually adds greater costs later when modifications are required. Therefore, part of the cost-benefit analysis should show how higher payouts in future years occur if certain requirements are overlooked in the initial design process.

- *Public works department representatives (for local government stations)* — Examine station design in terms of longevity and maintenance concerns.

- *Health and safety officer representatives* — Should be involved during a fire/EMS station design process. As most consensus standards such as NFPA are voluntary, a clear need must be demonstrated by the fire service to meet these requirements (perform cost-benefit analyses or show increased department liability).

- *Building and fire code officials* — Usually ask that the structure meet all applicable codes and ordinances, typically those found in the uniform building and fire codes.

- *Citizens groups* — Ensure that the station does not affect the local community in detrimental ways such as providing pedestrian safety hazards, noise, or an appearance inconsistent with the aesthetics of the neighborhood. It is important to consider including their participation in the building committee.

- *Consultants* — May be hired for any number of reasons to assist during the design process with some special need or to provide a general understanding to help department personnel who are not familiar with particulars of the station design process.

Task Analyses

The task analyses used to determine the types of activities firefighters perform can be applied to the planning of new stations and facilities (Figure 5.6). A task analysis for facilities might include apparatus maintenance, lawn care, materials storage, or cleaning and disinfecting tasks. Each of these tasks affects the design criteria being developed. From a larger perspective, the task analysis for a facility should address the actual use of the structure. Types of structures could include the following:

- Apparatus storage
- Occupied living quarters with apparatus bay
- Firefighter training
- Administration
- Combination facilities including the apparatus bay, living quarters, administrative functions, communications, or firefighter training
- Specialized training props

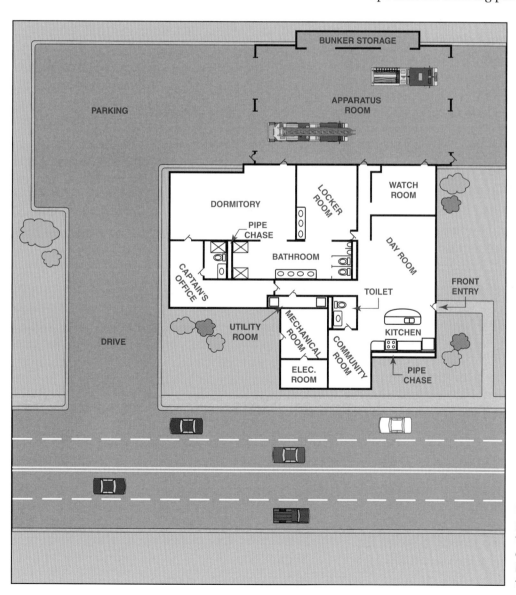

Figure 5.6 This sample floor plan of a new construction fire station includes a public community room and handicapped-accessible toilet.

- Apparatus maintenance
- Physical fitness training

The function of each type of structure places different health and safety requirements on the ultimate design.

Inspections

Another source for information are the inspections made in the current facilities. These inspections indicate a history of existing problems that can be eliminated by design changes in the new structure. Other sources for information are local or regional fire departments that have recently built new facilities. These fire departments provide additional points of view that are influenced by similar environments and demands.

Applicable Laws, Codes, and Standards

The health and safety officer is an advocate for safety and makes recommendations to the authority having jurisdiction. Recommendations are based on the applicable safety requirements of various codes, standards, and regulations. The health and safety officer works closely as a team player with building officials, architects, engineers, and other experts.

The health and safety officer must also take into consideration the national design and construction standards. Occupational Safety and Health Administration (OSHA) requirements, where applicable, must be consulted and adhered to. The requirements of the Americans with Disabilities Act (ADA) of 1992 has also altered the design of fire department facilities to allow greater access for persons with limited mobility, sight, or hearing. Because most fire department facilities are public properties, access to the public cannot be denied. This access is especially important for facilities that have public meeting rooms. Handicapped-accessible toilets must be provided in addition to general access. In administrative and training buildings, these requirements for accessibility are necessary for the nonemergency or nonuniformed employees of the department. Handicapped designated parking must also be provided under ADA requirements. It is important to recognize that not all areas of fire department facilities are open to the public. Telecommunications operations involving dispatching or primary or secondary public safety answering points may have limited access for the public. Storage and disinfecting areas may also be "off limits" to nondepartment personnel or civilians. These restrictions are based on local jurisdictional standard operating procedures or regulations.

A final resource for fire department facility design criteria is the United States Fire Administration (USFA) and the National Fire Academy (NFA). The USFA and the American Institute of Architects jointly produced and published a guide for the design of fire stations and facilities. Although it may no longer be in publication, it can be found in the resource center of the National Fire Academy along with other similar material. A more current resource is the federal Fire Administration's publication FA 168/May 1997, *Safety and Health Considerations for the Design of Fire and Emergency Medical Services Stations*.

Security Considerations

During the last 40 years of the twentieth century, the issue of security for public employees and facilities has become more important. Civil unrest, violence in the workplace, and terrorism have focused on government structures and personnel. In planning the design of new facilities, the health and safety officer must address this concern. Unfortunately, there are no published or mandated guidelines for mitigating situations of this nature. In order to develop design criteria to prevent such hostile acts, the health and safety officer needs to look at past incidents and consult with local law enforcement personnel. Postincident reports by the federal government following terrorist attacks on embassies or shootings within government buildings all contain recommendations for preventing future occurrences. Some security considerations that might be applicable include the following:

- Building set back from streets
- Metal detectors at public entrances
- Security lighting around structure
- Secure parking for employees
- Bulletproof glass in doors and windows
- Automatic door closers on overhead apparatus room doors
- Burglar/intruder alarm

Although not all suggestions found in the postincident reports are applicable, they are a good place to start research. For additional information, refer to NFPA 1221, *Standard for the Installation, Maintenance, and Use of Public Fire Service Communication Systems*. Another source of information may be the department's insurance carrier's loss control department. It may have ideas and checklists for security concerns that need to be addressed.

Renovation/Remodeling of Existing Structures

Most fire department facilities are built with an expectancy to stay in active use for forty to fifty years. However, in some urban areas of the United States and Canada, fire stations built at the turn of the century are still in active use. In rural areas, volunteer organizations still maintain the fire hall or station that was originally built to house horses and steam pumpers. Periodically, all fire department structures need renovation or remodeling. The health and safety officer must be a part of the planning process to see that existing hazards are removed and to ensure a safe work environment for the membership.

Resources available to the health and safety officer during the renovation/ remodeling process are the same as those used for the new station design process. The health and safety committee can provide input from the membership, the inspection reports can point to violations of codes, surveys of local departments can provide new insight, and a review of the task analysis for the department can point out the types of functions performed in the facility.

Upgrade to Laws, Codes, and Standards

Most model building codes in the United States and Canada require that the renovation of a space include an upgrade to current safety standards. It is very important to consult the building code in the jurisdiction to determine the minimum upgrade requirements. With the assistance of the department's fire prevention division, the health and safety officer can justify the upgrades by illustrating the safety benefits of the changes. It is important that the fire department facilities meet or exceed the safety standard requirements placed upon the general public. This builds credibility with the public and makes enforcement of fire codes a little easier.

Building upgrades do not involve just the safety-related issues such as means of egress, fire suppression, alarm systems, and heating, ventilating, and air-conditioning (HVAC) systems. The Americans with Disabilities Act of 1992 also places requirements on the fire department to provide access to public facilities for persons with disabilities. Disabilities include (but are not limited to) hearing, vision, and mobility. Accessibility includes designated parking spaces, braille instructions on placards at the entry doors, ramps or walkways that can be negotiated in a wheelchair, and more spacious bathrooms (Figure 5.7). The health and safety officer should make a thorough inspection survey of all fire department facilities for ADA compliance. In addition, the health and safety officer should note the impact that ADA-mandated changes may have on the design of the facilities. In some cases, this survey and impact information may help to have a facility exempted from ADA requirements. For instance, if a fire department has a two-story fire station with living quarters on the second floor and the public has no need to visit these quarters, it may not be necessary to make the second floor handicap accessible. This determination will have to be made by the local authority having jurisdiction. In any case, the health and safety officer must be prepared to document his recommendations for changes that fail to meet the ADA minimum requirements for accessibility.

Figure 5.7 Handicapped-accessible parking and entry doors must be provided for all department facilities.

Security Considerations

The issue of facility security is also important in the renovation/remodeling process. The level of security has to be determined by the potential threat to the structure, its openness to attack, and the cost of changes to be made. Again, existing government reports and recommendations from postincident analyses help guide the health and safety officer (Figure 5.8).

Community Expectations

Finally, changes in community expectations of the services provided by the fire department must be considered in the design plan. These expectations result in a need for a more flexible design — one intended to evolve with the changing needs of the community. Some questions that must be considered are as follows:

- Can the facility be used for other municipal functions?

- Will shared facilities meet the needs of the fire department?

- How many years will the facility be in use?

- What is the most cost-effective way to build the facility?

- How large should the facility be?

- What is the most effective spacing of new and existing facilities?

- How will community growth and demographic changes affect the facility location?

- Is it more cost-effective to renovate the facility or replace it with a new structure (Figure 5.9)?

In determining the design of new facilities or the renovation/remodeling of existing structures, the health and safety officer can use the risk management model outlined in Chapter 2, The Health and Safety Officer as a Risk Manager. This model provides a systematic approach to the planning process and maintains the level of documentation needed to justify recommendations (Figure 5.10).

Figure 5.9 The cost of renovation of older fire department facilities must be weighed against the cost of replacement. Not all older structures can be brought up to current building and fire code requirements.

Figure 5.8 In some areas, exterior doors to fire stations may be locked. An intercom system may be required so that visitors can request entrance.

Figure 5.10 The health and safety officer must be familiar with local codes that may require the installation of range hood suppression systems in all fire department stations.

Health Hazards

The health and safety officer deals with two categories of hazards within fire department facilities: health hazards (discussed in following sections) and physical hazards (discussed later in this chapter). Both types of hazards can be addressed through changes in structure design and through the education of personnel. Health hazard concerns include (but are not limited to) the disinfecting of contaminated clothing and equipment, indoor air pollution, water quality, proper personal hygiene, and infection control.

Designated Cleaning and Disinfecting Areas

Both OSHA Title 29 *CFR* 1910.1030 and NFPA standards 1500 and 1581 require new and existing fire stations to have two separate areas designated for cleaning and disinfecting. One area is designated for the cleaning of equipment and protective clothing and the other is for disinfecting emergency medical equipment. It is the responsibility of the health and safety officer or the infection control officer to designate these areas.

The health and safety officer or infection control officer should consult local health codes to determine the appropriate filtration, storage, or disposal of contaminated waste water from the cleaning and disinfecting areas. This determination should be done during the design phase of a new facility or during the installation of these areas in existing facilities. Cleaning materials used in these two areas, such as scrub brushes, disinfectant solutions, soaps, and sponges, should be stored within easy reach of the sink. To prevent further exposure or contamination, chemical goggles, aprons, and vinyl medical gloves must be provided by the department (Figure 5.11). The health and safety officer should provide a written protocol for disinfecting and cleaning of protective clothing, medical equipment, and fire fighting tools and apparatus. This protocol should be posted in the disinfecting area and included in the health and safety training for all personnel. The health and safety officer or the infection control officer must ensure that all personnel wear proper clothing when involved in cleaning or disinfecting activities. Equipment that has been cleaned or disinfected should be allowed

Figure 5.11 Designated EMS disinfecting sinks must be provided in fire department facilities that support EMS operations. PPE must be available in the disinfecting area.

to drain into the sink and not onto the floor. This prevents the possibility of tracking contaminated water into the living quarters and prevents a slipping hazard. Well-ventilated drying areas for clothing and equipment should be provided.

The disinfecting, cleaning, and drying of SCBA facepieces are extremely important. If they are not properly cleaned, SCBA facepieces can spread infection through the eyes, nose, and mouth of the wearer. Disinfecting and cleaning facepieces must follow the manufacturer's recommendation and the guidelines found in the current edition of NFPA 1981, *Standard on Open-Circuit Self-Contained Breathing Apparatus for the Fire Service*, or other applicable respiratory protection standards. The written protocol for cleaning and disinfecting should designate the location and method for maintaining the SCBA facepieces. However, the cleaning and disinfecting must *never* be done in the kitchen or bathroom of the fire station.

In addition to the cleaning and disinfecting sinks, an eyewash stand and disinfecting shower should be located within the work area or apparatus bay (Figure 5.12). In the event of a corrosive liquids spill, the unit must be available to the injured employee to reduce potential injuries through the flushing of the contact area or diluting of the corrosive liquid. The eyewash stand and disinfecting shower must have proper drainage and not dump directly onto the floor where contaminants can be tracked into the living quarters or create a slipping hazard.

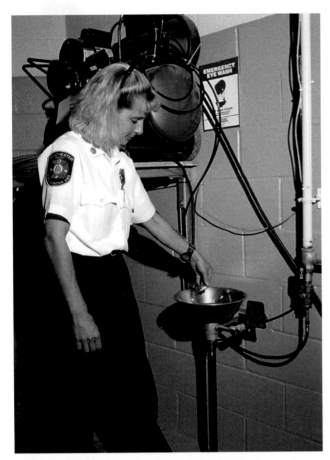

Figure 5.12 Eyewash stands must be provided in work areas where corrosive liquids are used or stored.

Designated Cleaning Area

The designated cleaning area, which should be located in the apparatus bay, must be equipped with proper ventilation, lighting, and drainage facilities. Drainage must be connected to the sanitary sewer system or septic tank and located away from the kitchen, sleeping and living quarters, bathroom, and designated disinfecting area. Personal protective clothing, fire fighting equipment, and portable equipment are cleaned in this area. The immediate cleaning of contaminated clothing and equipment prevents the potential contamination of other personnel or other portions of the facility.

Designated Disinfecting Area

The designated disinfecting area is primarily intended for the disinfecting of medical equipment exposed to bloodborne pathogens. It should be designated for disinfecting only and not used for the cleaning of mops, other materials used for general cleaning in the station, or equipment and clothing that have not been contaminated by blood or other body fluids. The area should be equipped with a stainless steel, two-bay utility sink with both hot and cold water service, nonporous rack shelving, adequate drainage, proper lighting, and proper ventilation. The disinfecting area should be located in the apparatus bay to encourage use immediately upon return to quarters. Immediate disinfecting of contaminated medical equipment reduces the possibility of exposure to bloodborne pathogens by other personnel.

Indoor Air Pollution

An environmental health issue of recent concern is indoor air pollution. Sources for this type of pollution can be external to the structure, internal mechanical systems or building materials (such as carpets), from the vehicles in the apparatus bay, or from the improper storage of chemicals or cleaning supplies. Indoor air pollution may also result from spray-painting operations in a mechanics bay.

External Pollution

External pollution from automotive traffic, local industry, or natural sources can only be controlled by keeping the pollution out of the structure. This means designing a structure that is totally self-contained without unfiltered exterior air entry. Older structures should be updated with storm windows (Figure 5.13), new air-handling systems, and increased weather stripping around openings. It should be noted, however, that completely sealed and self-contained structures can contribute to a sick-building syndrome. This syndrome is a condition in which more than 20 percent of the occupants suffer from adverse health effects. Symptoms may include headaches, allergic reactions, or fatigue and may be caused by gasses emitted by products used in the construction or renovation of the building.

Internal Pollution

Internal sources of air pollution include the heating and cooling system, water heater, use of tobacco products, and mold and mildew in the heating ducts (Figure 5.14). NFPA 1500 requires the installation of carbon monoxide (CO) detectors in both new and existing structures. These detectors provide early warnings if the levels of CO exceed safe limits. The health and safety officer should ensure

Figure 5.13 Station windows must be inspected to ensure that a complete thermal barrier is present and that external pollution cannot enter the facility.

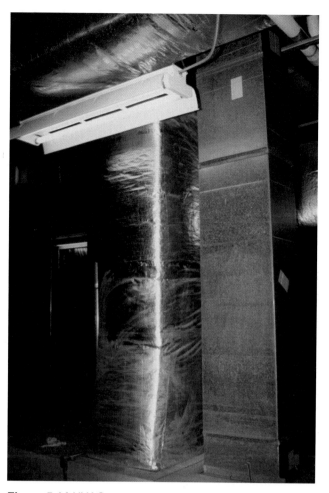

Figure 5.14 HVAC systems must be inspected and cleaned annually to prevent the growth of mildew and mold.

that these units are properly installed and operating. In addition, the health and safety officer should inspect the water heater, boiler, and mechanical systems for proper venting and complete combustion in the firebox. In some jurisdictions, annual inspections of these systems are mandated by outside agencies.

During new construction or renovation, air-handling systems should be designed to provide the proper rate of air exchange for the size of the facility. Building codes usually have the recommended minimum rates of exchange listed in tables based on the cubic feet per minute (cfm). Installing the correct size air-handling unit helps prevent the accumulation of stale air or carbon monoxide within the structure. Proper design of the ventilation duct system is critical.

During the inspection of the air-handling systems, attention should be given to the condition of the ducts. Dust accumulation, mildew and mold, and loose joints all can contribute to air pollution. This type of pollution can aggravate allergies and irritate nasal passages of employees. Ductwork should be inspected and cleaned periodically.

Finally, internal air pollution can result from the transmission of secondhand smoke from the consumption of tobacco products. NFPA 1500 mandates that smoke-free areas be provided in "work, sleeping, kitchen, and eating areas." Many departments have implemented nonsmoking policies that limit the areas where smoking is allowed.

Vehicles in Apparatus Bay

Another major source of air pollution is the apparatus and power equipment stored in the apparatus bay. When apparatus, light vehicles, generators, power tools, and lawn equipment are started, they create a great deal of carbon-based pollution. This type of contaminant builds up in a closed or unvented apparatus bay and can enter the living quarters through penetrations and air handlers. The health and safety officer can use both mechanical means and administrated policy to reduce this pollution.

Mechanical means include general room ventilation fans, point-of-capture apparatus ventilation systems, or mechanical roof vents. The most expensive and effective approach is the point-of-capture system that attaches to the apparatus exhaust pipe and detaches when the vehicle leaves the building (Figure 5.15). Use of a system of this type can remove major quantities of contaminants from the air. Unfortunately, it will not clear the air of pollution from the other power equipment operated in the apparatus bay. Therefore, a dual system that also includes the general room ventilation fans or mechanical vents is recommended (Figure 5.16). In the design of new structures, ventilation systems should be designed into the structure. Positive-pressure air systems within the living quarters that keep the transmission of apparatus room air out should also be built into the building.

The health and safety officer can implement administrative policies that require the opening of apparatus bay doors, the starting of fans before equipment is operated, and the wearing of proper approved respirators while operating equipment. Education helps to mitigate part of the air pollution problem around fire department facilities.

Water Quality

Water quality is a concern in most areas of the world. Water sources for fire department structures include municipal water distribution systems, wells, portable tanks, or cisterns. Sources of water contamination are industrial pollutants, agricultural pollutants, hard water, bacteria, and piping in the water distribution systems. The health and safety officer should ensure that the water used by the department personnel is clean and potable, regardless of the source of the water. This cleanliness would be particularly important where wells, cisterns, and portable tanks are used. Testing of the water should be done on a regular basis. If the source is a municipal water distribution system, then the health and safety officer should test the water for possible pollutants from the pipes within the structure. Iron pipes corrode over time and pollute the water they carry.

Where necessary, water filtration systems and water softeners should be installed within the facilities. These systems should be at the point where

Figure 5.15 Point-of-capture ventilation systems that attach to apparatus exhaust systems are the most efficient means of removing pollutants from the apparatus bay.

Figure 5.16 General room ventilation systems should be wired to operate when the apparatus room doors open or when engaged manually.

Diesel and Vehicle Exhaust Hazards

Diesel engines used in fire apparatus produce a mixture of toxic particulates and gases as the result of the combustion process. The composition of this exhaust product depends on several factors such as the specific fuel used, temperature of the engine, condition of the engine, and cleanliness of the air-intake filter among others. An analysis of general diesel exhaust has revealed a variety of extremely toxic substances at significant concentrations, including the following:

- *Oxides of nitrogen* — Any combustion in air produces various nitrogen oxides. Short-term exposures can cause respiratory tract irritation and infections. Long-term exposures result in lung-tissue damage and difficultly breathing.

- *Carbon monoxide* — It is produced as a by-product of combustion. Exposure to high levels of carbon monoxide causes death by tying up the hemoglobin in blood and preventing oxygen intake by the body. Exposure to carbon dioxide at lower concentrations causes headaches, dizziness, weakness, and neurological problems.

- *Volatile organic compounds (VOCs)* — These compounds are a class of carbon-based chemicals such as benzene, toluene, phenol, and chlorinated solvents. Many of these chemicals cause a variety of adverse health effects such as headaches, nausea, neurological disorders, respiratory irritation, and liver damage. Some VOCs are known or suspected carcinogens.

- *Polyaroma nuclear aroma (PNAs)* — PNAs are a class of relatively large, complex chemicals principally formed during combustion processes. In diesel exhaust, these chemicals often adhere to the soot particles. Most PNAs are documented carcinogens.

Much of diesel exhaust, including the smaller soot particles, is invisible. Therefore, exposure cannot always be detected. Furthermore, diesel exhaust can penetrate into clothing, furniture, and other items with which firefighters have routine contact. It can be later released after the initial exposure or absorb into firefighters' skins. Continued exposure to diesel fuel emissions has been linked to cancer and other serious health disorders.

Both the National Institute for Occupational Safety and Health (NIOSH) and OSHA have declared human exposure to diesel exhaust as a potential occupational carcinogenic (cancer causing) hazard through toxicological studies. A 1985 study commissioned by IAFF involved the measurement of diesel exhaust emissions at selected fire stations in New York, Boston, and Los Angeles. This study indicated that the most significant source of firefighter exposure to diesel exhaust was from the exhaust remaining in the station after the apparatus engine started. Some variations in the study results were identified, based on differences in climate, station design, number of runs per tour, and whether or not firefighters smoked.

---

the water enters the structure or at least enters the living quarters. Water heaters, water chillers, laundry washers, ice makers, bathing facilities, the kitchen, and the disinfecting sink should all rely on filtered water. Faucet-mounted filters can be used in stations containing iron pipe systems.

Proper Personal Hygiene

The area of personal hygiene is one of the most difficult areas of health hazards for the health and safety officer to handle. The health and safety officer must stress policies, training, and education in proper hygiene. Personal hygiene is a fundamental defense against the spread of germs, bacteria, and communicable diseases.

Policies regarding personal hygiene include taking sick leave rather than working when ill, not sharing protective clothing, cleaning and disinfecting SCBA facepieces following each wearing and according to the SCBA

manufacturer's recommendations, and washing hands before preparing food and after using the bathroom. Some departments are issuing individual sets of protective clothing and SCBA facepieces to help prevent the potential of cross-contamination. However, it is still possible for contamination to occur, so it is important to continue to emphasize that members must wear only their own protective clothing.

The health and safety officer ensures that new facilities and renovated facilities have the proper hygiene features to reduce the potential for infection. In existing facilities, features may have to be added periodically (as the budget allows) to improve infection control. Some of these features are as follows:

- All fire department facilities should have the proper hygiene equipment for wash... and disinfecting such as mop s... washers, and disinfecting and clea...

- Rest rooms and bathrooms should be designed and maintained so that they are not sources of infection.

Design features:
— Sinks and countertops should be nonporous materials.

— Walls, floors, and other surfaces should be constructed of materials such as tile or terrazzo that are nonporous and easily cleaned. Preferably, the material should not be conducive to slips when wet (Figure 5.17).

— Rest rooms and bathrooms should have push-to-open exit doors with no handles. Handles provide a place for infectious agents to accumulate and breed.

— Paddle-handle sink faucets do not require the user to grasp them to turn them on or off. In the event that an old-type faucet exists, the user should turn the faucet on when washing but use a paper towel to turn off the faucet after drying.

Figure 5.17 Daily cleaning of the station helps to ensure a ealthy living environment.

— Liquid soap dispensers should be installed rather than using the old-fashioned soap dishes that accumulate filth.

— Automatic flushing toilets and urinals should be considered in the design or renovation of fire department facilities. These types prevent the accumulation of waste that can occur with manual-type valves and does not require the use of hands for the operation of the valve.

— Hand-drying materials should be disposable or an air-drying machine should be available. These procedures decrease the possibility of infectious bacteria accumulating or breeding on multiple-use cloths. Although there is a possibility of germs or bacteria being on disposable hand-drying material, the possibility is remote.

Maintenance:
— Rest rooms must be properly ventilated.

— Antibacterial deodorant blocks should be provided for both types of toilet fixtures.

The reinforcement of good personal hygiene is very important. Hand washing is an important element in the process and should be strongly encouraged. In each rest room and bathroom a sign stating "**WASH YOUR HANDS**" should be prominently posted to remind members to take proper measures to protect themselves (Figure 5.18).

Laundry Facilities

A residential style clothes washer and dryer should be installed in each station. These machines should be located in a laundry room that is kept clean and orderly. In older facilities, it may be necessary to install them in a mechanical space, work area, basement, or apparatus room. Local codes should be consulted for proper location, draining, and venting. Only noncontaminated station/work uniforms and linens should be used in the designated washer and dryer (Figure 5.19). Protective clothing, medical use clothing, or contaminated station/work uniforms and materials should not be cleaned in these units. Contaminated clothing, items covered with smoke, grime, soot, grease, oil, or other nonbiological contamination must be cleaned in the industrial style washer provided for such use

Figure 5.18 The health and safety officer may find it necessary to post reminders to good hygiene such as this "wash hands" sign.

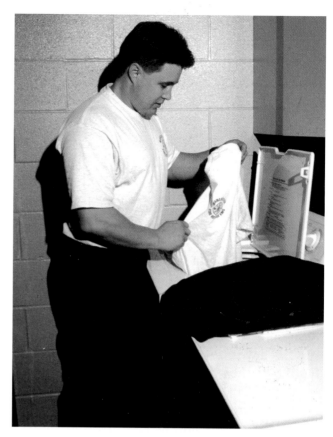

Figure 5.19 Laundry equipment should be supplied for the cleaning of linens and noncontaminated protective clothing.

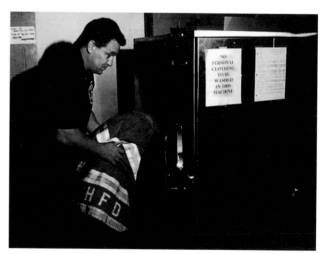

Figure 5.20 Industrial washers should be provided for the cleaning of protective clothing.

(Figure 5.20). Items of clothing that have been exposed to biological contamination must be disinfected in the area designated for that operation. Personnel are required to wash their station/work uniforms at the station. This prevents them from taking home their dirty and contaminated station/work uniforms and cross-contaminating family laundry. The fire department should provide all necessary detergents, proper water temperature, and laundering instructions at the fire station. Cleaning and disinfecting of station wear uniforms, personal protective garments, and structural fire fighting clothing are specified by Sections 3-5, 3-7, and 6-4.5 of NFPA 1581.

Industrial style washers and dryers should be provided in a central location for the cleaning of protective clothing, station work uniforms, salvage covers, or other potentially contaminated materials. These machines can be installed at district/ battalion stations, training centers, or supply buildings within the department. Personnel should follow specific cleaning instructions provided by the manufacturer of the protective clothing. Most protective clothing should not be dried in any type of dryer, and ventilated drying areas located out of direct sunlight should be provided.

Infection Control

Personal hygiene is a basic approach to infection control within the fire department. Infection control measures, like personal hygiene, include station design, policies, and education. The health and safety officer must consider the types of infection control needed in the following various areas:

- Rest rooms
- Kitchen
- Sleeping quarters
- Living/dining rooms
- Laundry, cleaning, and disinfecting areas
- Storage areas.

Rest Rooms

Rest rooms, as mentioned in the personal hygiene section, have one of the greatest potentials for infection if they are improperly designed or if members fail to practice proper hygiene or both. Policies regarding hand washing, proper use of disinfectant cleansers, and disposal of waste materials should be developed and posted as necessary.

Kitchen

The kitchen has the next greatest potential for the transmission of germs, bacteria, and communicable diseases. When planning a new facilities design, the health and safety officer should follow the health codes for commercial cooking establishments within the jurisdictional area. The guidelines for kitchen design are as follows:

- Countertops and sinks should be stainless steel with backsplash protection.
- Because of the potential for excessive use by a large number of people, commercial-grade appliances such as refrigerators, stove and ovens, and dishwashers should be installed if economically possible.
- Wooden chopping blocks or cutting boards should not be installed or used due to the potential for transmitting germs and food poisoning.
- Range ventilation hoods with fire suppression systems should be installed per code.
- A fixed fire suppression system is recommended because of the potential for inadvertently leaving cooking unattended.

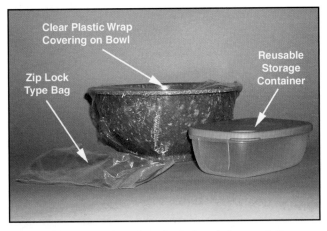

Figure 5.21 Examples of methods for saving partially consumed food in the refrigerator are plastic wrap, freezer bags, and resealable airtight containers.

- Clean, dry storage should be provided for all cookware, serving dishes, and food.

The health and safety officer should instruct all personnel in the proper handling and storage of food (Figure 5.21). If necessary, face masks and food preparation gloves should be supplied. Partially used containers of food and leftovers must be properly covered, sealed, and stored in the refrigerator or disposed of following meals. Because extremely hot water, usually 140° to 170°F (60° C to 76°C), is used to disinfect cooking utensils, personnel must be trained to wear rubber gloves to prevent burns. Keeping contaminated clothing or equipment out of the kitchen must be reinforced through education.

Sleeping Quarters

In the sleeping quarters, infection control can be maintained physically through proper ventilation and air exchange, separation of beds, installation of partitions between beds, and the issuance of individual sets of sheets, pillows, and blankets to all personnel. Sheets, blankets, and pillowcases must be washed daily if they are shared between shifts or personnel. They must be washed in separate loads from noncontaminated station wear or other work clothing. Policies should be in place to prevent personnel from bringing or wearing contaminated clothing into the area. Turnout boots and pants should not be brought into the sleeping quarters (Figure 5.22). These articles must be kept in the apparatus bay. Sleeping attire such as workout shorts and T-shirts can be required for nightclothes.

Figure 5.22 Protective clothing should be kept near the apparatus and not taken into the living area or sleeping quarters.

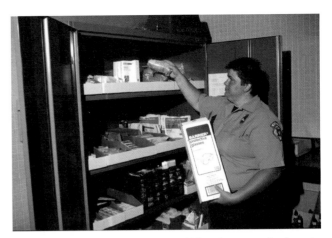

Figure 5.23 Supply storage should be clean, marked, and secure.

Living/Dining Areas

Living/dining areas may be the easiest areas in which to control potential infection. Air exchange rates, basic housekeeping techniques, and policies prohibiting contaminated clothing and smoking can reduce hazards in these areas. Dining surfaces should be maintained and kept clean. They should not be used for cleaning tools and equipment, disinfecting materials, or other dirty work.

Laundry, Cleaning, and Disinfecting Areas

Information regarding laundry, cleaning, and disinfecting areas may be found at the beginning of the section on Health Hazards and also under Laundry Facilities.

Storage Areas

Clean equipment and unused supplies should have a separate storage area. This separation helps to ensure that they are isolated from contaminants. The storage area needs to be conspicuously marked and secured (Figure 5.23). In the event a separate area is not available, a secure locker or cabinet can be used as a suitable substitution. Storage areas should be well ventilated and drained to prevent the growth of mildew and mold. Mechanical and electrical spaces should not be used for storage. Care should be used when storing flammable liquids. NFPA 30, *Flammable and Combustible Liquids Code,* should be consulted for proper storage of flammable liquids.

◆ Physical Hazards

Physical hazards include slipping, tripping, falling, injuries from basic housekeeping duties, poor internal and external lighting of facilities (illumination), noise pollution, electrical hazards, specific areas of hazard (shop areas, apparatus maintenance shops, apparatus bays, offices, and walkways), and hazardous building materials (asbestos). Once again, building design, training, and administrative policy can be used to mitigate these potential hazards. By far, the best option is to eliminate the potential hazard during the design phase and therefore remove the need for an administrative policy.

Slip, Trip, and Fall

To prevent slipping, tripping, or falling, floors should be kept clean, dry, and free of loose items or spills. Floors in the living quarters are usually terrazzo, tile, or smooth concrete. Although easy to clean and maintain, these materials have a slick surface that can contribute to slipping, especially when wet. In new construction or renovation, nonslip floor coverings should be considered.

The health and safety officer should ensure that the following guidelines are met to prevent slipping, tripping, or falling:

- **"Caution: wet floor"** signs should be provided for use during cleaning activities.

- Traffic patterns used to reach the apparatus bays must be kept free of obstructions.

- Doors must swing in the direction of travel where possible.

- Stairs must be equipped with handrails and non-slip treads.

- All walk areas must be well lit.

- Two-tier bunk beds must not be used due to the climbing and falling hazard inherent in their design.

- Apparatus bay floors should be well drained and nonslippery.

Many existing multistory fire stations are equipped with slide poles or chutes. In order to make these poles/chutes safer, they must be equipped with cushioned landing pads at the base. This procedure reduces the possibility of sprained or broken ankles. The upper-story entry point to the pole/slide should be enclosed in its own room or have a guardrail surrounding it (Figure 5.24). Weight-activated clamshell doors should be part of the design. Visitors and nonfire department personnel should be kept away and discouraged from using these devices. (**NOTE:** The trend for most fire departments is to avoid installing slide poles in new stations because of their inherent hazards. Some departments are instead installing slides where two-story stations are desired or necessary.)

Basic Housekeeping Procedures

Basic housekeeping procedures can help to control physical hazards in fire department facilities. Most departments adhere to a daily general station cleaning regimen and a weekly heavy cleaning day. During the daily cleaning, floors are swept and mopped, upper surfaces are dusted, materials are stored, linens and towels are washed, and kitchens and rest rooms are disinfected. Weekly cleaning takes care of the larger tasks such as cleaning apparatus room walls and ceilings, outside cleaning, washing windows, cleaning out cabinets and stor-

Figure 5.24 Slide poles in multistory stations must be protected with safety rails.

age areas, and stripping and waxing floors. Stripping and waxing floors, though done during this work period, should only be done annually or based on the floor manufacturer's instructions.

Illumination

Fire department facilities should be well lit internally and externally. When possible, windows and skylights should be used to let in normal daylight. Artificial light should be sufficient for most tasks such as reading, preparing reports, maintenance, or cooking (Figure 5.25). Glare, especially in office and training areas, should be kept to a minimum. Most structures use recessed fluorescent lighting fixtures to provide nonglare illumination. All fixtures should be kept clean with operating bulbs in place. Damaged fixture covers should be replaced to prevent the shattering of glass bulbs. Apparatus bay fixtures may be fluorescent, carbon-arc, mercury vapor, or incandescent. These should be in

Figure 5.25 Good illumination must be maintained through the replacement of burned out bulbs.

weatherproof fixtures and provided with protective light covers or cages. Exterior fixtures should be designed and maintained in a similar fashion.

Traditionally, fire station alarm notification has consisted of a combination of visual and audible systems. The combined effect of light and sound has been shown to add to the stress experienced by firefighters while the light tends to cause temporary night blindness. There is currently no standard or regulation to govern this type of notification system. Some departments, however, have analyzed the problem and developed alert systems that create less stress and do not affect the ability to see.

Noise Pollution

Noise pollution is generated by HVAC systems, machinery, apparatus, and office systems such as printers, telephones, and fax machines. Noise is amplified because of hard surfaces and noninsulated areas within the structure. Noise is a contributing factor in increased levels of stress and tension within the workplace and can contribute to hearing loss. It can interfere with speech communication, be a distraction to mental activities, and

be a general annoyance. Finally, it can increase work errors and decrease performance. The United States Government has established limits for exposure to noise based on the intensity of the noise and the duration of exposure. When the maximum exposure level is exceeded in the workplace, feasible administrative or engineering controls must be used. If such controls fail to reduce the sound levels, hearing protection must be provided and used to reduce the sound levels (Figure 5.26).

During the design phase, the health and safety officer should make every attempt to ensure that the following guidelines are met to reduce the potential level of sound within the structure.

• Mechanical spaces should be enclosed with soundproof materials.

• HVAC systems should be designed and purchased with quiet operation in mind.

• Ceiling tiles should be of sound-absorbing materials, and walls and floors should utilize soft surfaces where possible.

Figure 5.26 The health and safety officer must monitor the noise level of facility mechanical spaces to ensure that it does not exceed the acceptable level.

- Commercial grade or indoor/outdoor carpet should be considered because it can help to reduce noise within the building.

- Areas such as workout rooms or television rooms should be segregated from the sleeping area and study areas.

Electrical Safety

Electrical hazards are significant in fire stations due to the amount and use of water in the building. The building, electrical, and fire codes in effect in the jurisdiction must be followed in both the design and use of station electrical systems. As an example, most electrical codes require the installation of waterproof receptacles in areas such as in the apparatus bay and on the exterior of the structure where water may be present. In addition, electrical codes require the use of ground fault interrupter receptacles in kitchens and rest rooms. These devices reduce the shocking hazards associated with electricity in the presence of water. Storage in electrical spaces must be prohibited or controlled based on the local code. The health and safety officer should establish policies concerning the use of electrical tools and equipment based on the appropriate fire, building, electrical, or life safety code.

Fire stations should be equipped with auxiliary generators to provide sufficient electrical power to the facility for at least 24 hours (Figure 5.27). This system can be powered by natural gas, propane, diesel, or gasoline. Testing of the system under load conditions should be performed weekly to ensure proper performance. All personnel should be familiar with the operation of the backup electrical system.

Electrical Appliances, Extension Cords, and Outlets

Equipment such as portable fans and heaters should only be used in areas where they will not be knocked over or will not present a tripping or fire hazard. Electrical cords and telephone wires should be secured. All aisles should be free of storage items or materials that could cause a fall. Use of extension cords should follow the local fire and electrical code requirements. Installation of extra outlets is safer and easy when planning new stations. Adding additional outlets to existing buildings is a better solution (Figure 5.28). Extension cord use should be discouraged (Figure 5.29).

Figure 5.27 Stations must be equipped with auxiliary power generators to ensure constant power for essential communications equipment.

Figure 5.28 Overloaded receptacles constitute a hazard and should be an indication that additional receptacles are needed.

Figure 5.29 The use of portable space heaters and long extension cords should be discouraged.

Specific Areas of Hazard Concern

Specific function areas of the facility, such as the shop area, apparatus maintenance areas, apparatus bay, and offices, pose their own physical hazards. The following sections address each specific area of the fire department with hazard concerns.

Shop Areas

Shop and maintenance areas contain power tools and equipment that require care in their operation. The health and safety officer should ensure that all personnel are trained in the safe and proper use of the equipment. Hearing protection, goggles, and gloves should be provided in these areas. Personnel should inspect equipment regularly for defects or broken parts prior to usage and remove, repair, or replace defective or broken equipment. Depending upon the types of work conducted in the shop, ventilation is another concern that the health and safety officer should address. Personnel should keep the shop area clean and well lit. The fire department should provide hydraulic lifts, forklifts, or portable stairs for high stack storage of parts and materials and for replacing lights in apparatus bay fixtures (Figure 5.30).

Apparatus Maintenance Facility

If an apparatus maintenance facility is operated by the fire department rather than a private vendor, it must meet the building and fire code requirements for that type of operation. These code requirements include waste fluid and battery storage and disposal, paint storage, spray booth requirements, and other DOT, EPA, NFPA, and OSHA requirements. The shop area must be kept clean, free of tripping hazards, free of oil and fluid spills, and properly ventilated (Figure 5.31). According to EPA requirements, apparatus maintenance bay floor drains should be piped to an oil/water separator prior to discharge.

Apparatus Bay

In the apparatus bay, personnel should secure loose equipment to prevent tripping or falling. Point-of-capture exhaust systems should be kept attached to the vehicles or secured overhead out of the way. Walk paths should be properly marked and kept clear. Floors should be dry and drain covers kept in place. Apparatus doors should be equipped with

Figure 5.30 Forklifts can be used for storing materials on high shelves and racks. Qualified personnel should operate these units.

Figure 5.31 Maintenance areas in stations must be clean, ventilated, and well illuminated.

automatic safety features to prevent them from striking the apparatus or an individual (Figure 5.32). These safety controls may be mandatory on all new and existing overhead-type doors in some jurisdictions based on building code requirements.

Offices

Because the office is a nonemergency work environment, safety is often overlooked. Many office injuries are caused by an unsafe act by employees that involves tripping or falling. The health and safety officer should ensure that consideration is given to the type and quality of office furniture purchased. For example, chairs should be purchased that provide proper back support, are comfortable, and will not tip over. Personnel should be trained in the following office safety guidelines:

• Chairs should not be used as a ladder or a step.

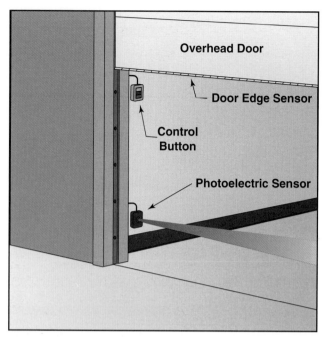

Figure 5.32 This illustrates the typical safety devices added to overhead doors. Door edge sensors cause the doors to reverse direction when they impact an object. Photoelectric cells prevent the door from operating if the beam of light is broken.

Figure 5.33 Files drawers in office areas may create a tripping hazard if left open.

• Exercise caution when working with filing cabinets. Drawers left open can cause a tripping injury or cause the file cabinet to tip over (Figure 5.33). File cabinets should be anchored to walls or floors to prevent tip over, especially in earthquake prone areas.

• Office supplies such as pens, pencils, and letter openers should be carefully stored with points down to prevent puncture wounds.

• When not in use, paper cutters should be kept with the blade down and should be equipped with a blade guard.

Kitchens

Physical injuries in kitchens are usually associated with cuts from sharp objects such as knives, can openers, and graters. These items should be properly stored to prevent cuts and abrasions. Personnel should be trained in the proper use of kitchen implements. Additional training in kitchen safety should be provided to reduce the potential for grease and steam burns.

Rest Rooms and Locker Rooms

Rest rooms and locker rooms, usually designed together, should be spacious to prevent crowding during shift changes. Personnel should keep locker doors closed except when in use. The department should provide permanently attached benches rather than loose chairs or stools and ensure that mirrors are permanently attached to walls and not freestanding. Wall-mounted clothing hooks should be above eye level.

Physical Fitness Areas

Physical fitness or workout spaces can create safety hazards as well. Good housekeeping is essential in a workout room. Personnel should maintain this area by performing the following:

• Clear the floor of loose objects.

• Maintain fitness equipment so that it is in good working order.

• Periodically clean equipment to prevent contamination from perspiration.

• Store loose weights in racks when not in use.

• Provide a storage area for exercise mats.

• Provide training in the proper use of equipment. Spotters or partners should be required when using free weights.

Mechanical Spaces

Mechanical spaces containing boilers, water heaters, and heating and cooling systems should be clean and well lit. No storage should be permitted in these areas. Ventilation should meet the existing code for these types of areas.

Exterior Areas

The health and safety officer should work with the appropriate personnel to ensure that the following exterior area guidelines are met:

- Exterior areas around the facility should be maintained and well groomed.

- Trash should be contained in an enclosed and secured area to prevent animals or children from gaining access to it.

- Parking areas should be well lit and, if necessary, secured by fencing and automatic gates.

- Garden edging, decorative plant materials, and rocks should be arranged to prevent tripping or falling. Slopes that are too steep to mow or maintain should be covered in low maintenance ground cover.

- Lawn sprinkler heads should be recessed where possible.

- Trees should be trimmed, and leaves should be removed regularly. Obstructions that could prevent motorists from seeing the apparatus leaving the station should be removed.

- Automatic traffic control devices or "Caution: Fire Station 150 feet (50 m)" signs should be installed on the street in front of the structure (Figure 5.34).

Walkways

Walkways, both interior and exterior, should be kept clear and free of loose debris. Railings must be provided on roof walkways and along paths where there is a grade change. All walkways must be lighted, and changes in grade must be visibly marked. Yellow tape or paint should be applied to steps and around work and storage areas (Figure 5.35).

Hazardous Building Materials (Asbestos)

Many older stations were built at a time when asbestos was used as a building material. This asbestos

Figure 5.34 Warning signs should be installed along streets to notify drivers of the location of fire stations and the possibility that apparatus may be entering the roadway.

Figure 5.35 Parking areas and walkways should be marked on apparatus room floors to reduce potential tripping hazards and clutter.

is most commonly found in floor and ceiling tiles and in pipe insulation. EPA and OSHA requirements have mandated the removal of asbestos from all public buildings. Removal of asbestos must comply with federal regulations and must be performed by a certified contractor. Some asbestos material can be encapsulated and left in place if it does not pose a threat to the occupant's health. This condition must be determined by the contractor hired to survey the facility.

◆ Ergonomics

Ergonomics is the applied science of equipment and workplace design intended to maximize productivity by reducing operator fatigue and discomfort. The health and safety officer must be aware of the science of ergonomics and how it relates to both station design and operational hazards. OSHA Title 29 *CFR* 1910.900 Subpart Y, *Ergonomics Program Standard*, applies to occupations in which the worker performs forceful lifting/lowering, pushing/pulling, or carrying activities. Ergonomic risk factors, also called *ergonomic stressors* and *ergonomic factors*, are elements of musculoskeletal disorder (MSD) hazards. *Musculoskeletal disorders* (MSDs) are injuries and disorders of the muscles, nerves, tendons, ligaments, joints, cartilage, and spinal discs. MSD hazards are physical work activities and/or physical work conditions in which ergonomic risk factors are present that are reasonably likely to cause or contribute to an MSD injury listed in Subpart Y. Ergonomic risk factors that pose a biomedical stress to the worker include the following:

- Force (forceful exertions, including dynamic motions)
- Repetition
- Awkward postures
- Static postures
- Contact stress
- Vibration
- Temperature extremes, both hot and cold

MSDs do not include injuries caused by slips, trips, falls, or other similar accidents. Examples of MSDs include the following:

- Carpal tunnel syndrome
- Rotator cuff syndrome
- De Quervain's disease
- Trigger finger
- Tarsal tunnel syndrome
- Sciatica
- Epicondylitis
- Tendinitis
- Raynaud's phenomenon or disease
- Carpet layer's knee

- Herniated spinal disc
- Low back pain

To address these MSDs and the resulting injuries, the health and safety officer should develop a full ergonomics program that includes the following six elements:

- Management leadership and employee participation
- Hazard information and reporting
- Job hazard analysis and control
- Training
- MSD management
- Program evaluation

Generally, the health and safety officer must consider issues such as workstations, tools, facilities, equipment, materials, and processes.

Engineering controls are the most effective means of eliminating MSDs before they occur (Figure 5.36). During facility construction or renovation phases, potential hazards created by storage height, equipment access, or office design can be addressed. Process or operational issues, such as proper lifting techniques, can be addressed through administrative policy and training.

Back injuries due to lifting can be addressed through the health and wellness program and training. In storage areas and apparatus bays, heavy

Figure 5.36 The health and safety officer should inspect workstations to ensure that proper ergonomic functions are met.

items should be stored at floor level. Shelves should be accessible without reaching or ladders should be provided. For moving heavy or oversized objects, task specific wheeled dollies should be provided or additional help obtained from personnel.

Regardless of whether or not the OSHA ergonomics program is mandatory in the jurisdiction, it is a model for a proactive approach to potential long-term injuries. The safety and health officer should consider applying it as a good practice to all aspects of the department.

 ## Safety During Natural Disasters

Natural disasters such as tornadoes, hurricanes, floods, and earthquakes can affect fire and emergency services facilities with the same potential for devastation as other parts of the community. The health and safety committee should design new facilities to withstand the highest level of disaster that is anticipated within the community. Some examples of design considerations are the following:

- Tornado safe rooms

- Earthquake-resistant structural members

- Facilities located outside of 100-year flood plains

- Adequate auxiliary power

The health and safety officer should ensure that personnel safety guidelines are written into the community disaster plan. If the frequency or severity of a disaster is high, medical, food, and potable water should be stockpiled at the site. Additional supplies are required if the site is designated as an evacuation center or relocation site for civilian victims. In any event, the health and safety officer should make sure that contingency plans are in place to provide personnel with the maximum amount of protection.

 ## Visitors

In most cases, fire department facilities are public buildings built with taxpayers' moneys. As such, they are open to the public and create an opportunity for the department to provide additional community service. Unfortunately, the trend to-

ward terrorism, vandalism, and workplace and domestic violence has made fire department buildings, along with other public structures, high-profile targets. To ensure employee and public safety, the health and safety officer, working with the administration, must develop a policy on public access to department facilities.

Whether the facility is a secured area or open to the public, personnel on duty should channel visitors through a check-in area. In administrative buildings, training centers, or other nonemergency facilities, visitors should report to a reception desk for processing. At emergency facilities, visitors should enter through the watch room. This area is usually staffed when companies are in quarters. A sign-in/sign-out process should be used to keep track of nondepartment personnel in the facility. This tracking would be especially useful when training facilities are used for evening or weekend classes for the public. A representative of the department should escort visitors when they are on department property to reduce the chances of theft or injury and to provide someone to give directions and answer questions.

As a community service, fire department facilities may be used for nondepartmental functions. Space can be made available for neighborhood meetings, CPR training, or as voting places among others. Classrooms, living spaces, or offices can be used for these purposes. The fire department should keep a record of the types of uses, the organizations using spaces, the frequency of use, and the number of attendees. This data can be used to justify the cost of the facilities to taxpayers and to maintain a positive public image.

Other permanent types of uses for department facilities can be as neighborhood medical centers for minor emergency care, police substations, household hazardous waste material collection sites, public utilities payment offices, or safe houses. A safe house provides the public, especially women and children, with a high visibility location to go to for protection (Figure 5.37). Fire department personnel should be trained in intervention techniques, symptoms of child and spousal abuse, and have a list of resources to contact. In multicultural neighborhoods, the fire department should either provide a translator with each company or provide a means

Figure 5.37 Fire stations may be used as safe houses for members of the community.

Figure 5.38 The health and safety officer should perform periodic inspections of all fire department facilities to ensure a safe working environment.

of access to a translator through an outside contact. If possible, bilingual fire department personnel should be assigned to specific neighborhoods where their talents can be utilized. Sign language training or an interpreter for the hearing impaired should also be provided to fire department personnel.

 Periodic Station Inspections

One of the duties of the health and safety officer is to perform both annual and monthly safety inspections of all fire department facilities. These inspections can be performed alone or in conjunction with the fire prevention bureau inspectors or the officer in charge of the facility (Figure 5.38). The inspections should target both physical problems with the structures and occupancy hazards created by their use. All code violations, repair needs, and deficiencies must be documented and reported

immediately. Records on each structure should be maintained for future reference. Sample inspection forms are included in Appendix G of this manual and in Appendix A of NFPA 1500 and Appendix B of NFPA 1521.

 Summary

The fire department health and safety officer is responsible for ensuring a safe working environment for department personnel. This responsibility can be accomplished through a proactive approach to new building design, safety improvements in renovated structures, policies on safe working conditions, and periodic facility inspections. Using the risk management model, the health and safety officer can develop a systematic, well-documented plan for addressing the safety hazards in the department's facilities.

Equipment Safety and Maintenance

Job Performance Requirements

This chapter provides information that will assist the reader in meeting the following job performance requirements from NFPA 1521, *Standard for Fire Department Safety Officer,* 1997 edition.

2-2.2 The health and safety officer shall have and maintain a knowledge of current applicable laws, codes, and standards regulating occupational safety and health to the fire service.

2-2.4 The health and safety officer shall have and maintain a knowledge of the current principles and techniques of safety management.

2-3.1 The health and safety officer shall have the responsibility to identify and cause correction of safety and health hazards.

2-3.2 The health and safety officer shall have the authority to cause immediate correction of situations that create an imminent hazard to members.

2-3.3 Where nonimminent hazards are identified, a health and safety officer shall develop actions to correct the situation within the administrative process of the fire department. The fire department health and safety officer shall have the authority to bring notice of such hazards to whomever has the ability to cause correction.

3-2.1 The health and safety officer shall develop, review, and revise rules, regulations, and standard operating procedures pertaining to the fire department occupational safety and health program. Based upon the directives and requirements of applicable laws, codes, and standards, the health and safety officer shall develop procedures that ensure compliance with these laws, codes and standards.

3-4.2 The accident prevention program shall provide instruction in safe work practices for all fire department members. This shall include safe work practices for emergency and nonemergency operations.

3-4.4 The health and safety officer shall periodically survey operations, procedures, equipment, and fire department facilities with regard to maintaining safe working practices and procedures. The health and safety officer shall report any recommendations to the fire chief or the fire chief's designated representative.

Chapter 6

Equipment Safety and Maintenance

The use of handheld tools, portable tools, and power tools plays an important role in fire service operations (Figure 6.1). The health and safety officer must ensure that the department's occupational health and safety program addresses equipment safety and maintenance policies. Safe use of these tools and equipment depends on both proper training in their use and proper maintenance and repair of the tools (Figure 6.2). The health and safety officer is responsible for implementing and managing these policies; however, he may delegate responsibility for actual compliance to station, training, shop, or maintenance supervisors and personnel.

Equipment safety and maintenance policies, protocol, and training can help to reduce the likelihood of personal injury or equipment damage due to the following actions:

• Improper use of tools and equipment

• Use of the wrong tool for the job

• Lack of care, testing, inspection, or maintenance of tools and equipment

• Poor housekeeping in shop or maintenance areas

• Lack of or improper use of protective equipment

• Improper use or storage of flammable or combustible liquids

• Lack of fire prevention procedures

Through the use of risk management and the task analysis, the health and safety officer can analyze the types of tools needed to accomplish the tasks of the department. He can then recommend

Figure 6.1 Power tools are used in many fire fighting operations including ventilation.

Figure 6.2 Safety must be considered when using tools to repair and maintain fire fighting equipment.

the proper tools for the jobs. He can also survey the maintenance program, determine potential hazards, and recommend changes. Finally, he can implement a training, testing, and inspection program that increases safety awareness and ensures proper maintenance of all tools and equipment (Figure 6.3).

Although the health and safety officer's primary guidelines for equipment safety are found in NFPA 1500, he should also be familiar with other NFPA standards that relate to fire service tools and equipment. These standards include, but are not restricted to, the following titles:

- NFPA 10, *Standard for Portable Fire Extinguishers*

- NFPA 30, *Flammable and Combustible Liquids Code*

- NFPA 33, *Standard for Spray Application Using Flammable or Combustible Materials*

- NFPA 51, *Standard for the Design and Installation of Oxygen-Fuel Gas Systems for Welding, Cutting, and Allied Processes*

- NFPA 55, *Standard for the Storage, Use and Handling of Compressed and Liquefied Gases in Portable Cylinders*

- NFPA 58, *Liquefied Petroleum Gas Code*

- NFPA 1932, *Standard on Use, Maintenance and Service Testing of Fire Department Ground Ladders*

Figure 6.3 Apparatus maintenance, such as the filling of apparatus tires, requires safety training.

- NFPA 1962, *Standard for the Care, Use and Service Testing of Fire Hose, Including Couplings and Nozzles*

- NFPA 1981, *Standard on Open-Circuit Self-Contained Breathing Apparatus for the Fire Service*

- NFPA 1983, *Standard on Fire Service Life Safety Rope and System Components*

In addition, the health and safety officer can rely on OSHA regulations for general industry guidelines for the use of equipment; IFSTA publications such as **Essentials of Fire Fighting** and **Principles of Vehicle Extrication** that include the proper use of equipment; and most importantly, the equipment manufacturer's guidelines for use, care, inspection, and maintenance. Materials referenced in NFPA standards and OSHA regulations, such as the American National Standards Institute (ANSI) standards, can be researched through the National Fire Academy or by contacting the appropriate standard writing organization.

The health and safety officer must take a proactive approach to safety issues involving the use, care, inspection, and testing of all tools and equipment. This chapter reviews in detail the following areas of concern:

- Shop safety

- Hand tool hazards

- Power tool utilization

- Portable power tools

- Manual forcible entry

- Rescue and utility ropes

- Portable equipment

- Breathing air compressors and cylinders (Figure 6.4)

- Flammable and combustible liquids

- Welding and cutting

- Fire prevention

 Apparatus and Equipment Maintenance Safety

Fire department maintenance can be performed either in a work area within the fire station or in a dedicated maintenance facility (Figure 6.5). The work can be done by uniformed personnel as part of their daily routine or by nonuniformed employees whose sole responsibility is equipment maintenance (see NFPA 1071, *Standard for Emergency Vehicle Technician Professional Qualifications*).

Figure 6.4 Only trained personnel should operate breathing air compressors and fill stations.

Figure 6.5 Apparatus maintenance personnel must wear eye protection while performing their tasks.

To ensure a safe working environment in the maintenance areas, the health and safety officer needs to develop policies that include requirements for personal protection, including the wearing of eye and face protection, hearing protection, and respiratory protection (Figure 6.6). These policies may also indicate training requirements and documentation of training in the use of these types of protection. For instance, the policy may require posting signs stating that eye, hearing, or respiratory protection is required while operating a certain tool or performing a certain task. Eye and face protection may consist of chemical goggles, safety glasses, faceshields, or equipment-mounted shields. Additionally, emergency eyewash sinks and showers should be provided in the work area in the event of an eye injury or exposure to caustic liquids.

Hearing Protection

Hearing protection can be reusable or disposable earplugs, earmuffs, or noise-absorbing materials

Figure 6.6 Shop personnel who operate apparatus must be familiar with the use of the communication headsets and the level of hearing protection that they provide.

to insulate the work area. Noise-level testing must be performed to determine the level of protection necessary based on OSHA exposure limits. A hearing conservation program should be part of the department's occupational health and safety program. A hearing conservation program includes the following components:

- Baseline audiometric testing upon employment with the fire department

- Annual audiometric testing and notification of results to the employee (Figure 6.7)

- Audiometric testing due to a significant noise exposure

- Noise monitoring and evaluation of all workstations/areas that could exceed permissible noise limits

- Mandatory use of hearing protection in work areas and on apparatus (Figure 6.8)

- Purchase of tools, equipment, and apparatus with noise-reduction devices

- Locating audible warning devices on apparatus and vehicles away from the crew compartment

- Proper posting of warning signs to others in the work area

The health and safety officer must evaluate various workstations for excessive or loud noise exposures. Regular monitoring of noise levels is imperative to ensure the success of this program (Figure 6.9). The health and safety officer must

Figure 6.8 Hearing protection must be provided at any workstation that has noise levels exceeding the specified OSHA exposure limits.

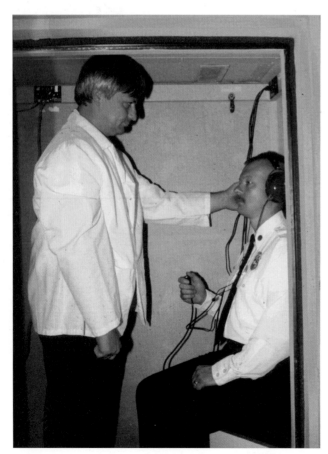

Figure 6.7 Annual audiometric testing is required to determine any significant hearing loss due to continued exposure to noise.

Figure 6.9 The health and safety officer monitors the noise level in work areas.

ensure that resources are available to conduct the periodic monitoring required for compliance with Title 29 *CFR* 1910.95, *Occupational Noise Exposure* (see *Hearing Conservation Amendment.*)

Respiratory Protection

Respiratory protection includes disposable filter masks, airline-connected masks, and cartridge-type respirators (Figure 6.10). Tasks such as sanding, sawing, or grinding that generate small quantities of solid particulates may only require the use of approved disposable filter masks. Tasks, such as large spray-painting jobs that create toxic gases require the use of either breathing airlines or cartridge-type respirators. The department should issue individual masks to maintenance personnel. Annual fit testing must also take place to ensure that the proper size mask is issued to each employee. Training and written program requirements for the department's respiratory protection program are found in Title 29 *CFR* 1910.134, *Respiratory Protection.*

Foot Protection

All personnel engaged in apparatus or equipment maintenance should wear steel-toed safety shoes. Some types of fire-fighting boots may meet the requirements of ANSI Z-41. However, the health and safety officer must refer to this ANSI standard and the manufacturer's literature to determine whether fire-fighting boots meet the standard and may be substituted for steel-toed safety shoes (Figure 6.11).

Safety Training

All personnel engaged in the maintenance of equipment and apparatus must be trained in the importance of safe work procedures, including the following:

- Wearing of eye, hearing, and respiratory protection

- Hazard recognition

- Proper tool usage

- Proper storage techniques

- Reporting damage and using proper repair procedures for tools and equipment

- Proper lifting and carrying techniques
- Proper general housekeeping practices
- Familiarity with materials safety data sheets (MSDS)

Access to the MSDS book is required by OSHA. The book must be updated when new products or materials are used in the workplace, and all employees must be familiar with the information maintained in the book.

Safe Work Environment

The potential for injuries can be reduced through safe work area design and cleanliness. Some hazards/exposures can be reduced during the engineering phase of building design and is the preferred method of risk reduction. Other safety hazards can be controlled by adding safeguards to

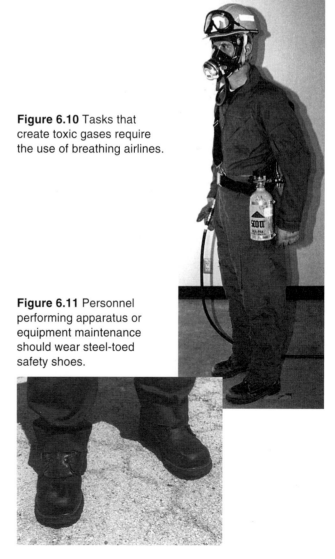

Figure 6.10 Tasks that create toxic gases require the use of breathing airlines.

Figure 6.11 Personnel performing apparatus or equipment maintenance should wear steel-toed safety shoes.

the individual equipment. Some of these design features include the following:

- Eye shields and mechanical safeguards installed on all power tools

- Large tools such as grinders and drill presses secured to the floor, workbench, or wall

- Noise-reducing materials installed on walls, floors, and ceilings where possible

- General room ventilation provided to filter and remove particulates and odors from the area

- Spray-paint booths provided in any area where spray painting takes place

- Rooms for the maintenance of SCBA regulators and oxygen regulators that are maintained according to OSHA regulations

- Adequate lighting; fixtures equipped with screens to prevent or contain lightbulb breakage

- Electrical receptacles equipped with ground fault interrupters (GFIs) to reduce shock hazards in the work area (Figure 6.12)

Potential hazards can also be controlled in the work and maintenance area through the following policies and procedures that promote cleanliness in the area.

- Keep maintenance areas clean and neat.

- Provide waste and trash containers and mark them according to their contents to ensure that solid waste is not contaminated with hazardous waste (Figure 6.13).

- Remove all waste from the work area daily and from the site at least weekly.

- Remove hazardous waste, oils, batteries, grease, etc., and dispose of them properly in accordance with OSHA and EPA requirements.

- Keep walkways clear to prevent tripping hazards.

- Clean up oil, grease, and liquid spills when they occur to prevent slipping and contamination of other areas (Figure 6.14).

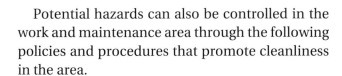

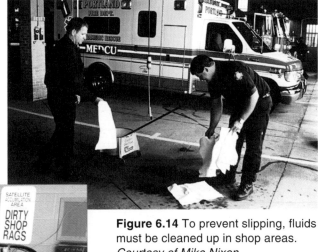

Figure 6.14 To prevent slipping, fluids must be cleaned up in shop areas. *Courtesy of Mike Nixon.*

Figure 6.12 Ground fault indicator electrical receptacles should be installed in all work areas where moisture may be present.

Figure 6.13 Personnel must properly dispose of contaminated shop towels in covered containers.

The shop supervisor or the health and safety officer must perform periodic inspections to ensure compliance with all safety policies.

Apparatus Maintenance Areas

Apparatus maintenance areas must be provided with appropriate lifting equipment for working on the apparatus. This equipment includes mechanical, electrical, pneumatic, and/or hydraulic jacks (Figure 6.15). In addition, personnel should address the following safety policies and procedures:

- Make periodic inspections of hoses, couplings, fittings, and joints to detect possible damage.

- Equip overhead lifts with fail-safe devices to prevent the apparatus from falling if power is lost.

- Follow the manufacturer's lubrication and maintenance schedules for the lifting equipment.

- Store jacks properly when they are not in use to prevent tripping hazards.

- Protect service pits by handrails or other means when the possibility of a person falling exists.

Figure 6.15 Manual lifting devices permit mechanics to position heavy or large items, such as tires, without injuring themselves.

 ## Hand Tool Hazards

Small hand tools are kept in the maintenance shop, training center, fire station, and on the fire apparatus. When not in use, hand tools must be stored properly. For safety and efficiency, they are kept in toolboxes, tool cribs, or on tool boards (Figure 6.16). Like all fire department property, they must be inspected and maintained periodically. Personnel should tag damaged tools and repair or replace them promptly (Figure 6.17). Tools that are issued to individual firefighters and shop maintenance personnel should also be repaired or replaced as necessary.

Training in the safe use of hand tools is part of the safety program. This training includes the proper way to use tools, select the correct tool for the task, and carry the tools. From a proactive approach, it should never be assumed that employees know the safe use of tools.

Figure 6.16 Hand tools must be stored and the work area maintained in a clean manner.

Figure 6.17 Tools that are damaged must be tagged and turned in for repair.

Nicked and Dull Blades

In general use, the cutting edges of blades become nicked and dull. Roofing knives, seat belt knives, and other cutting instruments should be sharpened according to the manufacturer's instructions. Personnel should use the following procedures:

- Discard tools that cannot be sharpened.

- Provide scabbards or sheaths to maintain sharp blades and prevent injuries.

- Regularly sharpen handsaws that are used for cutting wood or metal.

- Inspect and replace the tips of screwdrivers when they are damaged.

- Use machine oil to lubricate the metal parts of knives, saws, and screwdrivers to prevent rust and dulling.

- Inspect chisels and wedges for mushroomed heads, dull cutting edges, and nonsquare or chipped edges (Figure 6.18). Slightly rounded corners on the cutting edge help prevent breaking.

- Check chisels and wedges to ensure their reliability and compatibility with the material being cut.

- Wear adequate eye protection when using chisels and wedges to prevent injury due to chips and splinters. Make certain that personnel in the work area are shielded from flying chips.

- Wear heavy-duty work gloves while using chisels and wedges. Place a thick piece of sponge rubber over the head of the chisel to further protect the hand in case the hammer slips from the chisel head.

- When using a bull chisel held by one person and struck by another, hold the chisel with tongs.

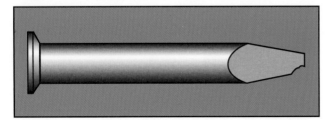

Figure 6.18 Chisels must be inspected for chipped blades and mushroomed heads.

Loose Heads

To prevent incidents involving loose heads on striking tools, the health and safety officer should ensure that the following repairs are made:

- Repair or replace loose heads on hammers, hand axes, or nail pullers (Figure 6.19).

- Replace cracked or splintered handles.

- Install safety grips on all types of hand tools for better grip and to prevent splinters.

◆ Power Tool Utilization

Power tools used in shop and maintenance areas include hand drills, sanders, saws, floor- or bench-mounted drills, lathes, table saws, and grinders. Training in the use of these tools should conform to the manufacturer's operator's manual. Knowledge of the operator's manual should also ensure that

Figure 6.19 Axe heads must be inspected for tightness to prevent the head from flying off.

Figure 6.20 Personnel operating drills and lathes must wear eye protection.

the tool is not used for the wrong task, which can lead to personal injury and damage to the tool.

Eye, hearing, and respiratory protection should be worn during the operation of all power tools (Figure 6.20). Hand protection is also advised for some power tools. Any guards, shields, or other tool-safety devices that the manufacturer provides with the power tool should be in place and used as intended. Removal or modification of safeguards can result in injuries, void the manufacturer's warranty, and expose the fire department to liability.

The health and safety officer must inspect all power tools regularly. Repairs must be made to ensure that tools are in safe working conditions. When working around flammable atmospheres, only electrical power tools, fans, lights, and other equipment having the Underwriters Laboratories (UL) listing or Factory Mutual (FM) approval for the specific hazard may be used. Electrical cords, plugs, and switches on power tools must be in good condition. The use of extension cords and multiplug receptacles must be in accordance with the local fire and electrical codes. All electrical power tools must be either double insulated or equipped with a three-prong grounded plug. Ground fault interrupter receptacles should be provided in the work area for all tools. In older facilities, this provision may require the addition of a dedicated circuit panel with breakers or the replacement of fuse boxes and wiring with current code-compliant products.

Good housekeeping around power tools is essential for safe and efficient operations and prevents tripping or slipping hazards, fire hazards, or other

Figure 6.21 Metal shavings must be cleaned up following the operation of metal cutting and grinding tools.

potential hazards that may be concealed. Loose debris, dust, metal shavings, sawdust, oil, and other waste material must be cleaned up following the operation of tools (Figure 6.21). Larger tools should be provided with a cover to reduce dust buildup.

 Portable Power Tools

Portable power tools in the fire service are used for shop and fire station maintenance and for emergency operations. The power source for these tools can be pneumatic (air-actuated or air-powered), hydraulic, electrical, powder-actuated, or gas-engine powered. Personnel assigned to operate these tools must be trained and qualified in their use. Proper supervision must also be provided during their use.

Personnel required to operate power-actuated tools must understand the potential for harm. Personnel must use all personal protective equipment provided by their department. Failure to comply with manufacturer's instructions and departmental procedures can cause serious injury.

Pneumatic Tools (Air-Actuated or Air-Powered)

Pneumatic (air-powered or air-actuated) tools located in shop areas are used for cleaning SCBA regulators and equipment and for vehicle maintenance. The air-powered tools are also used to fill tires and pressurize fire extinguishers. Air compressors or air-storage cylinders supply the air for these tools. Compressors and cylinders must be maintained in accordance with OSHA regulations and the manufacturer's instructions. To conduct the safest operations using compressed air, the health and safety officer must implement the following procedures:

- Equip all compressors for nonbreathing air with a safety relief valve in the event of overpressurization.

- Monitor water condensation in the compressor to prevent damage to tools and equipment and to prevent tank failure.

Figure 6.22 Only qualified and trained personnel may operate apparatus-mounted air compressors.

- Have an authorized inspector conduct a regular inspection and certification.

- Reduce the pressure of compressed-air nozzles used for cleaning purposes to less than 30 psi (206.8 kPa) and provide effective shields to prevent chips from injuring personnel. Permanently mark nozzles as meeting the 30-psi (206.8 kPa) requirement.

- Require that only trained and qualified personnel operate apparatus-mounted breathing air compressors (Figure 6.22).

Daily inspections and operations must include the following procedures:

- Inspect flexible hose for cuts, rotting, or other signs of deterioration.

- Inspect any tubing or fittings for leaks or signs of deterioration.

- Have authorized personnel perform any repairs, using approved parts.

- Wear eye protection when using compressed air.

Hydraulically Powered Tools

Hydraulically powered tools are usually used for cleaning apparatus and equipment, including fire hose and salvage covers. Water is the primary power source and comes from pressurized tanks, steam cleaners, and pressurized pumping systems. Operators must be instructed in the proper use of the tool, potential hazards, and correct protective clothing to wear (Figure 6.23). Improper use of steam cleaners can result in steam burns to operators or

Figure 6.23 Care must be used in operating high-pressure washers when used to clean contaminated hose and other equipment.

other personnel, damage to apparatus electrical systems, and removal of grease used for lubrication. Personnel should wear face protection, work gloves, and a waterproof jacket when using hydraulically powered tools.

Electrically Powered Tools

Electrically powered tools that do not have on/off switches or require constant pressure on them should have emergency stop switches. Constant-pressure switches must be designed and placed so that they can be easily and quickly used, yet not subject to accidental activation. Emergency switches must be located so that the operator can use the switch to shut off the equipment when working in any abnormal position.

Portable Power Tools Used at Emergency Incidents

Power tools used at emergency incidents are usually portable and powered from portable generators, air compressors, powder-actuated equipment, gasoline self-contained power units, air tanks, or apparatus-mounted generators. These types of tools include chain saws, circular saws, air-powered chisels, air bags, jackhammers, porta-power units, powered hydraulic extrication tools, and rope guns. NFPA 1936, *Standard on Powered Rescue Tool Systems*, provides the design, performance, testing, and certification criteria for these types of tools. Because the proper and efficient use of these types of tools can save lives and the improper use can cause injuries, operators must be trained, certified, and qualified where applicable (Figure 6.24). Peri-

odic training should be conducted to keep operators' skills in good order. Operators must be trained to recognize hazards created by operating power tools in potentially hazardous atmospheres (Figure 6.25). When not in use, tools must be secured on the apparatus to prevent damage or unauthorized use. This security is particularly important for powder-actuated equipment such as rope guns and ram set tools that use propellant charges.

Care and Maintenance of Power Tools

Tools must be cleaned, inspected, and operated on a regular basis and after each use. Repairs must be ordered immediately and the unit removed from service when repairs are required. When in operation, either during maintenance or at an incident, the proper protective equipment must be worn. This protection includes face/eye protection, gloves, and hearing protection. During emergency operations, full turnouts may be required, depending on the situation (Figure 6.26). To protect other personnel from loose debris, only those involved in the operation of the tool should be close to the operation. Other personnel must be moved out of range of flying hazards. When personnel cannot be moved back, salvage covers can be used to reduce the effect of flying materials. Salvage covers are also recommended for the protection of trapped victims.

Figure 6.26 Improperly used power extrication tools can injure personnel engaged in a rescue operation. Full PPE is required when power extrication tools are employed.

Figure 6.24 Serious injuries can result from the improper use of power tools during ventilation operations.

Figure 6.25 Air-operated lifting bags are essential for rapid extrication of victims trapped in overturned cars. Operators must be aware of potential hazards.

Manual Forcible Entry Tools

The category of manual forcible entry tools includes the following traditional types of fire service tools:

- Flat and pick-head axes
- Sledgehammers
- Pike poles
- Halligan tools
- Pry bars
- Ram-type door openers
- Crowbars
- Miscellaneous hand tools

All forcible entry tools should be secured on the apparatus to prevent the potential of flying objects in a vehicle collision (Figure 6.27). Tools should be secured with tool clips inside storage compartments, not on the exterior or in the crew compartment of the apparatus. Pointed tips, found on pike poles and pick-head axes, should be covered with tip protectors (Figure 6.28).

Training in the proper use of manual forcible entry tools is an essential part of the recruit training program (Figure 6.29). The health and safety officer

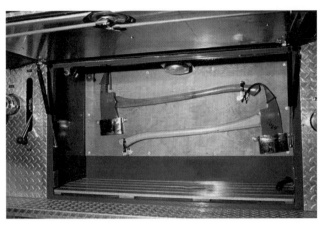

Figure 6.28 Pick-head axes properly stored in apparatus compartments have blades and tips covered.

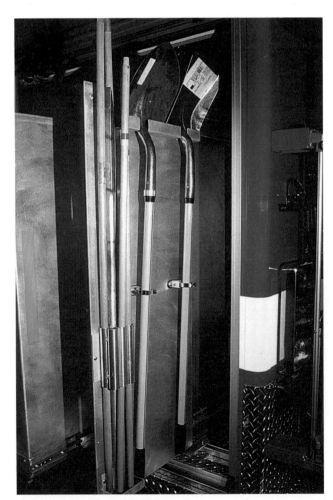

Figure 6.27 To prevent injury to personnel and damage to equipment, tools must be secured in compartments.

Figure 6.29 Personnel must be trained in the proper method for carrying axes and other sharp tools.

should take advantage of this concentrated training to include tool-safety training and the concepts of ergonomics.

Because forcible entry tools must be ready for use immediately, the health and safety officer must ensure that tools are inspected and maintained properly by having personnel use the following procedures:

• Inspect tools periodically and following each use.

• Clean tools to remove debris and oil coatings on metal parts to prevent rusting. The use of oil on all-metal tools should be based on the manufacturer's recommendations, type of tool, and local climate.

• Do not paint metal parts because paint can conceal damage.

• Varnish wood handles.

• Do not paint wood handles because it may cover damage (splinters or cracks).

Figure 6.30 Firefighters must be able to recognize damage to hand tools including loose heads, dull cutting edges, and split handles.

• Sharpen cutting edges of axes to manufacturer's specifications by using a hand file; the cutting edges should not be painted.

An example of maintenance procedures for the axe (the most common fire fighting tool) would be the following:

• Check wooden handles for cracks, blisters, or splinters in the wood.

• Check to ensure that heads are on tight (Figure 6.30).

• Check to ensure that cutting edges are free of nicks, burrs, chips, or tears.

• Check for rust, paint, or debris on heads and handles.

Further inspection, care, and maintenance information for forcible entry tools can be found in the IFSTA **Essentials of Fire Fighting** manual.

The health and safety officer should be aware that fiberglass handles are currently replacing wooden handles on some tools, including axes, pike poles, and sledgehammers. Fiberglass is lighter, more flexible, and less likely to produce splinters (Figure 6.31). However, it can break under stress and will shatter, producing extremely dangerous shards and splinters. Fiberglass handles require the same inspection and care as wood handles. It should not be painted because paint will cover potential damage.

 Rescue and Utility Ropes

Ropes are used to perform many tasks in the fire service. Some of these include the following:

Figure 6.31 Fiberglass tools require the same level of inspection and care as tools with wooden handles.

- Rescue

- Lifelines (Figure 6.32)

- Lifting and hoisting

- Fireguard lines

- Securing objects in place

Each task requires a specific type of rope, and all ropes require continual inspection and maintenance. The health and safety officer must analyze the tasks and recommend the correct type of rope for each task (Figure 6.33). He must also develop the written protocols for the testing, inspection, repair, cleaning, and disposal of ropes in use by the department. Through the training center, he must ensure that all members understand the protocols and the proper uses of the various types of ropes.

Figure 6.32 Lifelines or rescue ropes are used for the removal of victims from inaccessible locations.

Written Protocol on the Use, Inspection, and Care of Life-Safety Rope

- Use only approved life-safety rope for supporting personnel when ascending or descending.

- Life-safety rope must not be used for any purpose other than rescue operations.

- Life-safety rope must meet the standards set forth in NFPA 1983, *Standard on Fire Service Life Safety Rope and System Components*.

- Life-safety rope should have a minimum working load rating of 300 pounds (136.1 kg) for a one-person rope and 600 pounds (272.2 kg) for a two-person line.

- The minimum diameter of a one-person life-safety rope shall be ⅜ inch (9.5 mm).

- Once life-safety rope has been used in an emergency, it must be removed from service, destroyed, or relegated to use as a utility rope per NFPA 1500.

- Life-safety rope must be stored in a separate bag or container marked to indicate its use.

- All life-safety rope must be inventoried, and records must be maintained on its purchase, inspection, use, and disposal.

- Life-safety rope must be inspected periodically for cuts, abrasion, mildew, mold, or acid damage.

- The results of inspections shall be reported to the health and safety officer upon completion.

Figure 6.33 Utility ropes are used to hoist tools to upper floors or roofs of buildings.

Rescue Ropes

Rescue ropes, harnesses, and accessories, including life belts, must be inspected annually according to NFPA standards. The equipment must be marked, inventoried, and its use tracked from purchase to disposal. This responsibility may be given to the training officer, supply officer, or to the health and safety officer, depending on the department's organization (Figure 6.34). Guidelines for the selection, inspection, and use of rescue ropes and accessories can be found in NFPA 1983. The manufacturer's recommendations should be followed regarding cleaning and storage of rescue ropes.

Utility Ropes

Utility ropes used for hoisting, securing, and limiting access to areas do not require the level of inspection and care that rescue ropes do. They do, however, require periodic inspection, testing, cleaning, and maintenance. Utility rope should be inspected and cleaned following each use (Figure 6.35). Ropes that have been exposed to excessive heat, direct flame contact, or hazardous materials should be removed from service. Water can rot rope and cause pockets of mildew and mold to form within the strands. Ropes should be dried in well-ventilated spaces out of direct sunlight. Ultraviolet light may also contribute to the aging of rope fibers.

Storage

All ropes, utility or rescue, must be stored in waterproof bags and secured on the apparatus to prevent mechanical damage. The ropes should not be stored in compartments where oil or gas vapors are present. The bags must be marked with the length and diameter of the rope, type of construction material, and intended use (Figure 6.36). All rope should be inventoried and maintained (including the date and source of purchase and the date of disposal). This inventory assists in the decision whether to assign the rope to surplus or retire it.

 ## Portable Equipment

Portable equipment includes those items that are carried on the fire apparatus but are not attached to or part of the apparatus (Figure 6.37). This type of equipment includes the following items:

Figure 6.34 The health and safety officer must ensure that all personnel are instructed in the inspection, care, storage, and use of lifelines or rescue rope.

Figure 6.35 Utility rope must be cleaned after use.

Figure 6.36 Rescue rope and utility rope must be stored in marked bags in dry compartments on the apparatus. *Courtesy of Mike Nixon.*

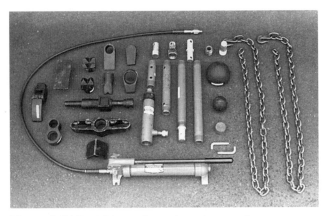

Figure 6.37 Portable equipment, such as porta-power units, includes those items that are carried on the apparatus but are not part of the apparatus.

- Ground ladders
- Fire hose
- Nozzles, accessories, and appliances
- Portable fire extinguishers
- Electric cord reels and lights
- Portable fans used for negative and positive-pressure ventilation
- Self-contained breathing apparatus
- Portable generators

With the exception of SCBA units, NFPA 1500 requires that portable equipment be inspected weekly and within 24 hours of each use. (See SCBA section for SCBA requirements.) In addition, the appropriate NFPA standards for ground ladders, hose, and extinguishers specify acceptance tests and annual in-service tests. Acceptance testing occurs when the department receives new equipment. Annual testing occurs on or near the anniversary date of the original acceptance test. Hose and ground ladder testing may be coordinated with the annual apparatus pump testing or with scheduled training events.

Ground Ladders

Although the acceptance testing of ground ladders must be performed by a certified tester or other qualified personnel, trained fire department personnel can perform the required annual testing of all equipment. This annual testing requires either the construction or purchase of test equipment. Depending on the department's budget, it may be more cost-effective to have the annual testing per-

Figure 6.38 The health and safety officer must ensure that ground ladders are inspected annually.

formed by an outside testing laboratory. The NFPA ground ladder standard requires that test results be retained on file. Ground ladders that do not pass the test must be removed from service and repaired. Ground ladders that cannot be repaired must be destroyed rather than sold. This eliminates the potential for liability. After ground ladders are used, personnel should make the following checks (Figure 6.38):

- Condition of rope halyards
- Indication of heat damage to beams, tips, and rungs
- Condition of rungs (loose, bent, or broken)
- Warpage of beams
- Operation of dogs, pawls, or locks
- Presence of debris, water, or ice that might impair the operation of the ladder

Gasoline-Powered Equipment

Gasoline-powered equipment such as generators, positive-pressure ventilation fans, or extrication tools must be inspected for mechanical damage, wear, rust, and proper operation (Figure 6.39). Damage must be reported immediately and the unit removed from service.

Self-Contained Breathing Apparatus

Self-contained breathing apparatus shall be inspected, tested, and maintained in accordance with NFPA 1500 and 1981. The department is required to have an SCBA program that meets the guidelines of NFPA 1500. Company personnel are responsible for the inspection and cleaning of SCBAs assigned to them (Figure 6.40). Inspection and cleaning are done daily and after each use. Factory certified personnel, who may be members of the department, members of another department within the jurisdiction, or outside independent contractors, must perform annual SCBA inspections, testing, and maintenance. Certified SCBA technicians must ensure that all SCBAs owned by the department are properly marked and inventoried. They must also maintain testing and repair information on each SCBA unit for its service life.

SCBA cylinders are regulated by the government. Specific requirements include the following:

- Hydrostatic testing of SCBA cylinders must be performed in accordance with the manufacturer's instructions and the guidelines of the applicable governmental agencies.

- Damaged cylinders and valve stems must be repaired or replaced.

- Cylinders must be tested in accordance with DOT regulations and retired after 15 years of service.

Additional information may be found in the current IFSTA **Self-Contained Breathing Apparatus** manual or in the IFSTA **Respiratory Protection** manual that will be published in late 2001.

Figure 6.39 Hydraulic hoses and connections on power extrication tools must be inspected regularly.

Figure 6.40 Firefighters must be trained in the proper inspection methods for SCBA.

Breathing Air Compressors and Cylinders

Air compressors and cylinders perform numerous functions within the fire service. As previously discussed, nonbreathing air compressors are used in shop and maintenance areas for cleaning, filling tires, and operating pneumatic tools and equipment. They also provide air for portable air-powered tools and equipment. Compressors that generate breathing air are used to fill individual SCBA cylinders, cascade systems, and high capacity air cylinders on aerial devices, rescue apparatus, hazardous materials units, and mobile and portable breathing air units.

Breathing Air Compressors

Breathing air compressor systems consisting of the compressor, piping manifold, cascade cylinder system, and fill station may be located in a fixed site or mounted in a mobile unit. Either way, the system must meet strict regulatory standards due to the life safety factor involved with breathing air. Some of the requirements for breathing air systems are as follows:

- Breathing air compressors must meet the operational and design requirements of national standards such as OSHA, NIOSH, Compressed Gas Association (CGA), and NFPA (Figure 6.41).

Figure 6.41 Breathing air compressors, cascades, and fill stations must be installed according to manufacturer's specifications.

- Compressors must be located in well-ventilated areas. They must be isolated from toxic atmospheres, including apparatus exhausts and general air pollution.

- Personnel operating air compressors must be trained in their operation. Operating instructions must be posted near the fill station controls.

- Hearing protection must be provided and worn when the breathing air equipment is in operation.

- Air receivers (cylinders) must meet American Society of Mechanical Engineers (A.S.M.E.) Boilers and Pressure Vessel Code Section VIII requirements.

- Manufacturer's instructions must be followed in the installation, operation, inspection, testing, and maintenance of breathing air systems.

- Air filters must be replaced at least periodically, based on hours of operation.

- Breathing air cascade cylinders must be secured to walls or floors in accordance with OSHA and NIOSH regulations and the local fire and building codes. Cylinders under pressure are extremely dangerous and must be handled with care.

- The manufacturer's recommendations for filling, inspecting, hydrostatic testing, and maintenance of breathing air cascade cylinders must be followed.

The breathing air system must be tested for purity (Figure 6.42). Requirements for this testing are the following:

Figure 6.42 The health and safety officer is responsible for the sampling and testing of the breathing air used by the department.

- Breathing air must be sampled at least quarterly per NFPA 1500 and tested from the transfer point from the system to the SCBA cylinders. Inspection of the compressor and cascade system and the sampling of the air may be performed by a department member, usually the health and safety officer.

- The health and safety officer draws the intake air of breathing air compressors and provides it to a third-party testing laboratory to ensure that it meets manufacturers' requirements.

- A third-party testing laboratory must be used to perform the testing of the air sample. The results of the air quality testing must be maintained as part of the SCBA program.

- Breathing air cylinders rented or purchased from an outside vendor must also be tested quarterly for air quality or the vendor must certify the air quality.

Additional breathing air requirements may be found in NFPA 1500 and 1981. These standards are more strident than those found in the CGA documents.

Use and Storage of Oxygen

The majority of fire departments that use medical grade oxygen for EMS operations purchase it from medical gas suppliers. In some instances, larger departments may fill their own individual cylinders from a bulk storage facility. Either way, the department must follow the regulations and guidelines for the use and storage of oxygen. Oxygen cylinders used by the fire department must meet the requirements of Title 49 *CFR* Part 178, *Shipping Container Specifications Regulations of the Department of Transportation.* Specific requirements include the following:

- All personnel must be trained in the use of the oxygen cylinders.

- Smoking must be prohibited during filling operations and use.

- Oily rags and gloves must not be used or stored around oxygen cylinders.

- Cylinders that are empty should be marked or tagged "EMPTY" (Figure 6.43).

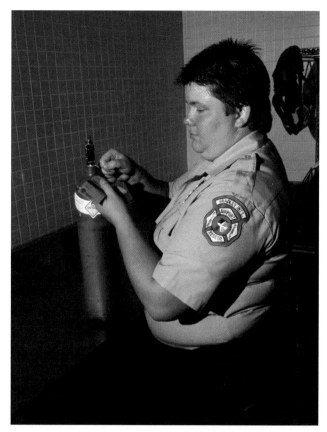

Figure 6.43 Empty oxygen bottles must be tagged "EMPTY" and stored for filling.

- Nonmedical oxygen used for cutting and welding equipment must meet the same Title 49 *CFR* Part 178 requirements.

- Cylinder valve caps must be in place when hoses are not connected to cylinders and during transport.

Flammable and Combustible Liquids

Flammable and combustible liquids of all descriptions and quantities are found in most fire department facilities. Because the hazards of flammable and combustible liquids are well known to the fire service, their uses, storage requirements, and fire prevention procedures are well documented. The health and safety officer in conjunction with the training branch and hazardous materials branch can perform a survey of the types of materials used within the department. From this survey, the health and safety officer can create a flammable and combustible liquids policy for the department. The policy should include guidelines for dispensing, disposal, storage, documentation of the materials, types of containers, and use (Figure 6.44). Training in flammable and combustible liquids hazards should be part of the firefighter's basic training program and should be repeated periodically during the firefighter's career.

Dispensing of Flammable/ Combustible Liquids

Training in proper dispensing techniques is essential for all personnel. Dispensing of flammable liquids can occur on a daily basis with the fueling of apparatus, refilling of gasoline-powered tools, and maintaining apparatus (Figure 6.45). The health and safety officer must consult the local fire pre-vention code in addition to the NFPA standards for correct procedures. The health and safety officer ensures that the following procedures are used during dispensing of flammable liquids:

- Use proper grounding of apparatus and fuel trucks during fueling.

- Prohibit smoking during the handling of flammable liquids (Figure 6.46).

- Use drip pans to contain spills of flammable liquids (Figure 6.47).

- Locate fire extinguishers in close proximity during fueling.

- Perform proper clean up of spills

Figure 6.45 Apparatus must be properly grounded during refueling operations.

Figure 6.44 Flammable and combustible liquids must be stored in NFPA-compliant safety cans with self-closing lids.

Figure 6.47 Drip pans must be located under dispensing taps on drums of flammable and combustible liquids.

Figure 6.46 "No Smoking" signs must be posted at fuel sites.

Storage of Flammable/Combustible Liquids

Local fire codes regulate the quantity of flammable and combustible liquids that can be stored at a particular facility and the type of storage that must be used. Small quantities of materials such as cleaning solutions and spray paint can be stored in commercial storage lockers or cabinets. Larger quantities of materials such as paint thinner, naphtha, and gasoline must be stored in UL-listed or FM-approved flammable and combustible liquids cabinets (Figure 6.48). Cabinets must be in well-ventilated areas of the facility and outside the living quarters. Warning signs must be posted in the flammable and combustible liquids storage area, and smoking must be forbidden.

Large capacity portable containers used for flammable and combustible liquids such as drums of gasoline, kerosene, solvents, or hydraulic fluids must be secured to the structure during storage. Depending on the local code and size of the container, a containment area capable of holding the liquid in the event of a spill may be required (Figure 6.49). Movement of portable containers must meet applicable codes.

Bulk storage of flammable and combustible liquids on fire department premises must meet the local building code, fire prevention code, NFPA standards, OSHA regulations, and other prevailing regulations (Figure 6.50). These regulations include

Figure 6.49 Large-capacity containment areas for containers used for flammable and combustible liquids may be required by state or local codes.

Figure 6.48 Fire department facilities must be provided with UL- or FM- approved flammable and combustible liquids storage cabinets.

Figure 6.50 Bulk storage of flammable and combustible liquids must meet the local building and fire codes as well as NFPA standards and OSHA regulations.

not only the storage and containment requirements but also the pumping, fire suppression, and handling requirements.

Flammable and combustible liquids containers must meet the requirements of NFPA 30 and local fire prevention codes. Self-closing UL- or FM-approved safety cans must be used for storage of gasoline, kerosene, and diesel. Military-style fuel cans (sometimes referred to as "Jerry" cans) or other containers without self-closing devices must be avoided. All containers must be marked with the names of the contents.

Documentation of Materials

All flammable and combustible liquids maintained in the facility must be inventoried and have material safety data sheets (MSDS) collected on each item. This information must be kept in an MSDS book on the site, and it must contain health, safety, and injury prevention information (Figure 6.51). Personnel must be trained in the use of the flammable and combustible liquids and the MSDS book. Materials must be used for their intended purposes and not for nonapproved uses. For instance, health concerns and regulations mandate that kerosene may no longer be used for apparatus cleaning.

Appropriate Fire Extinguishers for Flammable /Combustible Liquids

All fire department facilities must meet the requirements of the local fire prevention code, part of

which regulates the number, locations, and types of portable fire extinguishers. Further requirements can be found in NFPA 10, OSHA regulations; and other flammable and combustible liquids standards such as NFPA 30. The health and safety officer must review the survey of flammable and combustible liquids in use in the facility and determine the appropriate extinguishers for the individual hazards. The extinguishers must be visible, marked as to type, and properly mounted within the structure. Portable fire extinguishers carried on an apparatus may not be considered part of the units used for the protection of the structure.

Welding and Cutting

Welding and cutting operations occur in both emergency and nonemergency situations in the fire service. In emergency operations, cutting torches are used to gain entrance to a structure, remove structural debris in a building collapse, and free victims of a vehicle collision or structural collapse. Welding and cutting operations are used in maintenance facilities to repair damage to apparatus, construct training equipment, and build storage racks for hose and equipment. Both welding and cutting are very specialized tasks and require that operators be trained and qualified to perform them. Both the health and safety officer and the training branch should make certain that operators have met the required level of training.

Equipment used in welding and cutting operations must be in good condition (Figure 6.52). Periodic inspections of hoses, cables, fittings, cou-

Figure 6.51 The health and safety officer must instruct all personnel in the use of the MSDS book and advise them of any additions or deletions.

Figure 6.52 Cutting and welding torches must be inspected regularly.

plings, and tips must be made to ensure that equipment is in safe-operating condition.

Nonemergency Operations

In the maintenance facility, at the training center, and in the fire station, welding and cutting operations must follow the guidelines found in the appropriate regulations and standards. Some of these guidelines are as follows:

- Screens and faceshields must be used when necessary (Figure 6.53).

- Full protective clothing, including welder's helmet, leather gloves, and goggles, must be worn.

- The area must be free of potential fire hazards, loose debris, and flammable and combustible liquids and gases.

- Appropriate ventilation must be provided for the work area to decrease the potential of ignition of welding gases.

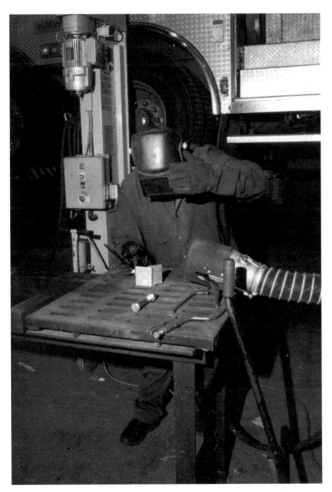

Figure 6.53 Welding faceshield, gloves, and protective coat must be worn during welding operations.

- Gas levels must be monitored during the welding operation.

- Power sources can include electricity or compressed gas.

- Electrical equipment must be grounded, and power cables must be protected from damage.

- Gas cylinders must be secured to the structure or in a portable cart.

- Cylinder caps must be in use on spare cylinders. Cylinders must meet the requirements of 49 *CFR* Part 178.

Fire extinguishers of the correct type for the hazard must be mounted in the work area. The local fire prevention code may require that a permit be issued for cutting or welding operations. The health and safety officer must consult this code for guidance.

Welding operations produce electrical arcs and gas flames that emit ultraviolet and infrared rays that can be extremely harmful to human skin and eyes (Figure 6.54). Permanent eye damage can result from looking at the arc or flame. The eyes and skin should be protected against the glare and radiation from a welding arc or flame. Helmets and hand shields should be used except where the arc or flame is completely contained (Figure 6.55).

Proper ventilation of the welding area must be provided. Welding produces toxic gases that can be harmful to personnel, and gases are also flammable.

Figure 6.54 Cutting and welding torches emit ultraviolet and infrared rays that can be harmful to the eyes.

Figure 6.55 Welder's gloves are essential for personnel to use when welding.

Figure 6.56 Firefighters with charged hoselines are ready to protect personnel engaged in a rescue operation. *Courtesy of Martin Grube.*

Emergency Operations

During emergency operations, operators must be equipped with the correct protective clothing, which includes full protective coat and trousers, boots, helmet, faceshield and goggles or safety glasses, hearing protection, and gloves. Where victims are present, protection must be provided for them. Salvage covers can be used to shield them from flying sparks and debris.

Before cutting begins, the area must be inspected for potential hazards that might be created from the cutting process. Flammable and combustible liquids and explosive atmospheres should be monitored and removed prior to cutting. If the hazard cannot be removed, another extrication process must be used. Charged hoselines or extinguishers must be in place in case of explosion or fire (Figure 6.56).

◆ Fire Prevention

The health and safety officer, in coordination with the fire prevention bureau, should develop an in-ternal fire prevention program. This program helps to ensure a fire-safe work environment and acts as an example to the public of the commitment of the fire department to fire prevention.

The fire prevention program should include facility and equipment inspections, fire prevention training, and a proactive approach to fire protection systems (Figure 6.57). The basis for the program should be the local fire prevention code or NFPA 101®, *Life Safety® Code*. It should include the following components:

- Fire prevention training for all personnel

- Portable fire extinguisher program, including inspection and installation

- Good housekeeping in and around facilities

- Installation of exit signs, emergency lighting units, smoke detectors, and other safety-related equipment (Figure 6.58)

Figure 6.57 The health and safety officer must ensure that compressed gas storage cylinders are capped and secured to the structure.

- Installation, maintenance, and inspection of fire suppression systems in facilities, range hoods, and paint booths

- Posting and enforcement of no-smoking signs where flammable and combustible liquids are in use

- Regulation in the use of portable space heaters

- Carbon monoxide (CO) detectors (Figure 6.59)

- Gas detectors

- Testing and maintenance of fixed fire suppression systems

The consistent enforcement of the fire prevention program can have the dual benefit of a safer workplace and instilling a proactive attitude among employees. In any case, the fire department should attempt to meet a higher standard of compliance than it expects from the public.

Figure 6.58 Emergency medical kits and eyewash liquids are part of the safety equipment that must be accessible in fire department facilities.

Figure 6.59 Smoke detectors and CO monitors must be inspected for proper operation.

◆ Summary

The health and safety officer plays an integral role in developing, implementing, and managing the occupational safety and health program. Equipment safety is an important part of this overall program. As with all portions of the program, a proactive approach is needed to ensure a safe work environment. This safe environment is achieved by applying the risk management model to determine the safety needs; developing, implementing, and managing the safety program; and thorough training of all personnel.

7

Personal Protective Equipment

Job Performance Requirements

This chapter provides information that will assist the reader in meeting the following job performance requirements from NFPA 1521, *Standard for Fire Department Safety Officer,* 1997 edition.

3-7.1 The health and safety officer shall review specifications for new apparatus, equipment, protective clothing, and protective equipment for compliance with the applicable safety standards, including the provisions of Chapters 4 and 5 of NFPA 1500, *Standard on Fire Department Occupational Safety and Health Program.*

3-7.2 The health and safety officer shall assist and make recommendations regarding the evaluation of new equipment and its acceptance or approval by the fire department in accordance with the applicable provisions of Chapter 4 of NFPA 1500, *Standard on Fire Department Occupational Safety and Health Program.*

3-7.3 The health and safety officer shall assist and make recommendations regarding the service testing of apparatus and equipment to determine its suitability for continued service and in accordance with Chapter 4 of NFPA 1500, *Standard on Fire Department Occupational Safety and Health Program.*

3-7.4 The health and safety officer shall develop, implement, and maintain a protective clothing and protective equipment program that will meet the requirements of Chapter 5 of NFPA 1500, *Standard on Fire Department Occupational Safety and Health Program*, and provide for the periodic inspection and evaluation of all protective clothing and equipment to determine its suitability for continued service.

Chapter 7

Personal Protective Equipment

Probably no area of the fire service has seen as much change in the last quarter of the twentieth century as the area of personal protective equipment (PPE). This change has included the development of new fire-resistant materials, new NFPA standards and OSHA requirements, and the increasing hazards encountered at emergency incidents. It is the responsibility of the health and safety officer to keep up with improvements in personal protective equipment, changes in safety standards, and trends in fire-fighting techniques. The health and safety officer must enforce the applicable sections of the OSHA regulations and NFPA standards.

Personal protective equipment provides protection for the eyes, face, head, body, ears, feet, hands, and respiratory system. Personal protective equipment also includes the personal alert safety system (PASS), auxiliary safety systems, and miscellaneous safety-related apparel (Figure 7.1).

Firefighters wear protective clothing or equipment when performing the following operations:

- Structural fire fighting
- Proximity fire fighting (Figure 7.2)
- Emergency medical incidents (Figure 7.3)

Figure 7.1 Full personal protective equipment for structural fire fighting includes helmet, flashover hood, gloves, boots, coat, pants, PASS device, and SCBA.

Figure 7.2 Incidents involving high levels of conductive, convective, or radiant heat require the use of proximity suits during extinguishments.

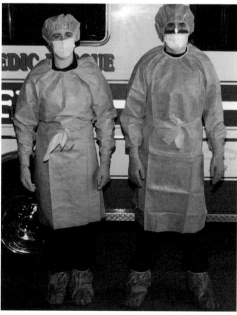

Figure 7.3 EMS incidents that pose a threat of bloodborne pathogens require personal protective clothing certified for that type of hazard. *Courtesy of Mike Nixon.*

- Wildland fire fighting (Figure 7.4)
- Confined space or structural collapse rescue (Figure 7.5)
- Special operations
- Hazardous materials incidents
- Station activities (Figure 7.6)

This equipment is intended to protect firefighters while allowing them to complete their duties. Modern PPE reduces the severity/frequency of injuries and permits firefighters to work in an otherwise hostile environment longer than when wearing previous types of protection. Because the firefighter's duties vary, the type, style, construction, and design of the equipment also varies (Figure 7.7). It is the health and safety officer's responsibility to determine the specific equipment required to accomplish the functions of the department. This responsibility can be accomplished through the risk management model discussed in Chapter 2, The Health and Safety Officer as a Risk Manager. However, personal protective equipment is useless if it is not used properly. It is the responsibility of the health and safety officer as well as the department management team to reinforce the need to wear the necessary clothing and equipment required by the hazards encountered.

Once the health and safety officer has determined the hazards that the firefighters face, he must consult NFPA 1500, *Standard on Fire Department Occupational Safety and Health Program*, and OSHA regulations Title 29 *CFR* 1910.132, *Personal Protective Equipment*, and Title 29 *CFR* 1910.134 *Respiratory Protection* to determine the level of protection required. With this information, recommendations for the types of personal protective equipment needed can be provided to the fire department administration.

Because very few fire departments have the financial resources to provide full sets of protective equipment for every employee and every possible hazard, the health and safety officer has to be prepared to recommend equipment based on the following factors:

Figure 7.4 Wildland personal protective equipment is designed to provide protection while remaining light in weight and allowing the wearer maneuverability.

Figure 7.5 Personal protective equipment intended for confined space use must not prevent the wearer from entering and exiting tight spaces.

Figure 7.6 Station wear may be constructed of fire-retardant fabrics or pure cotton and must not contribute to burns when exposed to flame.

- *Feasibility* — If the department rarely responds to certain types of incidents, specialized clothing and equipment may not be justified, and structural protective clothing and equipment may meet normal needs. Therefore, the department may not be able to provide some specialized clothing and equipment.

- *Cost* — Protective clothing that has the protective ratings established by NFPA, yet costs less than top-of-the-line clothing, may be acceptable (NFPA standards are minimum requirements).

- *Frequency of use* — Departments that do not have high numbers of incidents may be able to keep equipment in service longer than recommended or provide only enough clothing for the number of personnel on duty at a given time. Before making the decision to provide one set of clothing for multiple persons, the health and safety officer must consider the issues of hygiene and proper fit. Providing one set of clothing may prove to be false economy.

- *Maintenance requirements* — A design that does not require as much maintenance may be preferable to one that is complex.

- *Training requirements* — A design that employees are familiar with can reduce training time and costs.

Selection of the appropriate type of protective equipment is a complicated process and requires that the health and safety officer be prepared to weigh the difference between the maximum level of protection available and the level that the department can afford to purchase (Figure 7.8). The mandate, however, is to provide the minimum level as defined by state and federal regulations. Anything less is not allowed; anything more is a benefit. This consideration should be part of all ongoing risk versus benefit process (Figure 7.9).

NFPA 1521, *Standard for Fire Department Safety Officer*, mandates that the health and safety officer "develop, implement, and main-

Figure 7.7 High-angle rescue operations require specific clothing and life safety harnesses.

Figure 7.8 Level A personal protective equipment provides the highest level of protection to hazardous materials incidents. Use of this equipment is restricted to personnel trained in its use.

Figure 7.9 Level B personal protective equipment is the minimum level recommended for initial site entries until the hazards have been defined by monitoring and sampling.

tain" a protective clothing program as specified in Chapter 5 of NFPA 1500. A detailed description of this chapter is found in Appendix A of NFPA 1500. In addition, a step-by-step process for designing a personal protective equipment program can be obtained from the Canadian Occupational Health and Safety Office (Figure 7.10).

While NFPA 1500 describes the types of equipment to be used for specific incidents, other standards describe the design, testing, labeling, and certification of the equipment. The health and safety officer must be familiar with these standards in order to judge the value of the equipment he is recommending for purchase. The applicable standards include the following titles:

- NFPA 1971, *Standard on Protective Ensemble for Structural Fire Fighting*

- NFPA 1975, *Standard on Station/Work Uniforms for Fire and Emergency Services*

- NFPA 1976, *Standard on Protective Clothing for Proximity Fire Fighting*

- NFPA 1977, *Standard on Protective Clothing and Equipment for Wildland Fire Fighting*

- NFPA 1981, *Standard on Open-Circuit Self-Contained Breathing Apparatus for the Fire Service*

- NFPA 1982, *Standard on Personal Alert Safety System* (PASS)

- NFPA 1983, *Standard on Fire Service Life Safety Rope and System Components*

Figure 7.10 Some Canadian personal protective equipment differs in design and construction from that used in the United States.

- NFPA 1991, *Standard on Vapor-Protective Ensembles for Hazardous Materials Emergencies*

- NFPA 1992, *Standard on Liquid Splash-Protective Clothing for Hazardous Materials Emergencies*

- *NFPA 1994, *Standard on Protection Ensembles for Chemical or Biological Terrorism Agents* (*Publication date May 2001)*

- NFPA 1999, *Standard on Protective Clothing for Emergency Medical Operations*

An understanding of these design and testing standards gives the health and safety officer the ability to develop compliant specifications. Additionally, it helps determine whether the equipment he is evaluating meets the minimum requirements of the applicable standard.

This chapter provides the health and safety officer with an overview of the various types of personal protective equipment and the design, care, and maintenance of that equipment.

◆ Respiratory Protection

In the fire service, respiratory protection includes personal protection from toxic or hazardous atmospheres found at incidents, concentrations of dust, paint, and other particulates produced during cleaning and maintenance, and airborne microorganisms that may be present during medical emergencies. Each type of situation requires a different level of personal protection and a different type of equipment. Self-contained breathing apparatus (SCBA) can provide protection against all of these circumstances but are generally used to protect against toxic and hazardous atmospheres and oxygen-deficient atmospheres at emergency incidents. NIOSH-certified dust or particle masks provide the minimum level of protection and can be used in the shop environment. N-95 or high efficiency particulate air (HEPA) filter-equipped masks can be used in the shop environment as well as at medical emergencies. All masks or respiratory protection must meet the minimum requirements for the environment that the employee is in and be able to protect against the atmospheric concentration of the contaminant present (Figure 7.11).

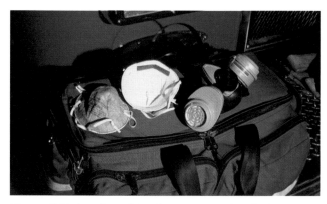

Figure 7.11 A variety of respiratory protection devices are normally used in the fire service.

Respiratory Protection Policy

A written policy covering the proper use of all respiratory protection equipment is required by NFPA 1500 and 1582 and by Title 29 *CFR* 1910.134. This policy can be developed by the health and safety officer or by the health and safety committee. When adopted by the department, it must be communicated to the members of the department. Like all operating procedures, it must remain a living document, be reviewed periodically, and be revised as necessary.

A respiratory protection policy requires the following elements:

• Selection criteria for respirators
• Written program
• Inspection criteria
• Proper use procedures
• Individual fit testing for employees
• Maintenance program
• Training program
• Air-quality testing program
• Medical evaluation and certification for those wearing respirators
• Procedures for regularly evaluating program effectiveness

The health and safety committee or any other designated body within the fire department can establish selection criteria for the purchase of respiratory protection equipment. The health and safety officer should be part of the selection team to ensure that respiratory protection safety requirements are met in the selection criteria. Considerations in the selection process are as follows:

• Types of hazards in which the department expects to operate
• Cost of purchase
• Periodic testing and maintenance requirements
• Training requirements
• Design requirements of NFPA 1981

Equipment used in fire fighting must be NIOSH certified for structural fire fighting. Particle masks and other types of respirators must meet the minimum requirements set by OSHA and be NIOSH certified.

Inspection programs for SCBA are established in NFPA 1500. Fire company personnel perform daily inspections of the SCBA unit at the beginning of the work period and after each use (Figure 7.12). Specific steps in the inspection procedure may be found in the IFSTA **Self-Contained Breathing Apparatus** manual or the **Fire Service Respiratory Protection Program** manual, which is due for release in 2001, and in the NFPA Selection, Care, and Maintenance (SCAM) documents for the SCBA standard (NFPA 1981). Units that are not in daily use must be inspected at least weekly. Monthly inspections include a check of the entire unit for deteriorated components, air tightness of the cylinder and valves, gauge comparison, reducing valve and bypass valve operation, and a check of the regulator, exhalation valve, and low-air alarm. Following each inspection (daily, weekly, or monthly), the unit is cleaned before returning it to service.

Figure 7.12 At the beginning of each work shift, company personnel are responsible for inspecting the SCBA assigned to them.

Inspection/maintenance personnel must be certified by the manufacturer to work on the specific type of SCBA owned by the department. The fire department may assign this annual inspection, testing, and maintenance to an internal inspection/maintenance function, rely on a separate department of the municipality, or contract the function to a private vendor (Figure 7.13). These annual inspections/testing involve all components of the system. The health and safety officer must be aware that the United States Department of Transportation requires the periodic hydrostatic testing of cylinders. The department must maintain a record of hydrostatic tests throughout the life of the cylinders (Figure 7.14).

Fit Testing

Individual facepiece fit testing is required of all employees who are going to use respiratory protection equipment (Figure 7.15). Fit testing helps to

Figure 7.13 A manufacturer certified technician must perform annual inspections of respiratory breathing equipment. *Courtesy of Weis American Fire Equipment Company.*

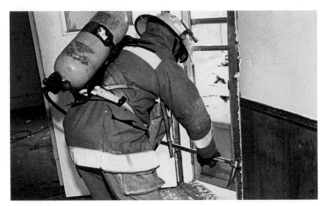

Figure 7.14 SCBA cylinders, like the one worn in this photo, must be hydrostatically tested and the resulting data maintained throughout the life of the cylinder.

Figure 7.15 A full and complete seal of the SCBA facepiece is ensured by fit testing personnel.

ensure a full and complete seal of the mask to the face. Testing must be done before allowing any employee to use the mask in a hazardous atmosphere. The requirement for a proper fit prohibits the wearing of any facial hair such as beards or sideburns that would prohibit a complete seal of the facepiece on the wearer. Specific steps in the fit testing process are found in the IFSTA **Self-Contained Breathing Apparatus** manual and in OSHA Title 29 *CFR* 1910.134. The health and safety officer may be assigned this testing activity and its documentation. Medical clearance is required before fit testing.

Maintenance Program

Certified technicians must perform the maintenance for respiratory protection equipment. Economics and efficiency determine whether the department uses internal technicians, technicians from other departments, or contractors from outside the department. Noncertified personnel must not be allowed to work on or make modifications to SCBAs. The department standard operating procedure (SOP) must include criteria for SCBA maintenance, repair referral process, tracking, and record keeping.

Training

All employees must be trained in the use of the respiratory protection equipment. Training includes the following guidelines:

• When to wear the equipment

- How to don and doff the equipment
- How to operate the equipment
- How to perform daily inspection procedures
- How to clean and maintain the equipment

This training is not a one-time event but must be an ongoing process. The health and safety officer may find it necessary to work with the training officer in providing this training.

Air-quality testing was addressed in Chapter 5, Facilities Safety, while the requirements for medical testing were covered in Chapter 3, Physical Fitness and Wellness Considerations. Both are essential parts of the respiratory protection policy.

Creation of Database

A final and basic part of the respiratory protection program is the creation and maintenance of a database. This database includes all documentation on the equipment and the personnel who use it (Figure 7.16). Because this data may be compiled by a variety of personnel inside and outside the department, the health and safety officer is the logical person to ensure that the necessary information is gathered and maintained in the appropriate file. The following information must be maintained:

- Personnel files, medical
  — Fit testing documentation

Figure 7.16 The health and safety officer is responsible for ensuring that data is collected, analyzed, and maintained on the respiratory protection program.

— Annual medical exam information regarding pulmonary tests
— Exposure reports
— Injury reports that involve respiratory injuries

- Personnel files, general
  — Fit testing, pass or fail
  — Documentation and required SCBA training

- SCBA and other respiratory equipment data
  — Design and selection criteria
  — Annual testing and inspection documentation
  — Maintenance records on individual units
  — Air-quality reports

- Monthly inspection
  — Hydrostatic test dates on cylinders

The health and safety officer must remember at all times that the individual medical records of department personnel are confidential.

◆ Personal Protective Clothing

Personal protective clothing is designed to NFPA standards and tested and certified by a third-party testing organization to protect firefighters from specific types of hazards. Certification is sometimes provided for a single garment such as a pair of fire-fighting gloves and sometimes for an entire ensemble such as an aircraft fire-fighting entry suit. Therefore, the health and safety officer must be aware of the specific use that a garment or ensemble is certified for when purchasing clothing (Figure 7.17). Personal protective clothing should be

Figure 7.17 The use of full personal protective equipment is required from initial attack to salvage and overhaul operations because of the hazardous products of combustion present in structure fires.

specified and purchased based on the individual department's intended use of the clothing. NFPA standards provide the minimum protection requirements for each of the primary functions. Some garments may have dual certification for more than one function. An example of a dual-certified garment would be one that protects the wearer from the hazards of a structural fire incident and also from biological hazards encountered in an EMS incident. If the health and safety officer determines that dual-certified garments are to be specified, purchased, and issued, he must consult the appropriate standards. In addition, the health and safety officer should consider the effects of climate when selecting personal protective clothing. Personal protective clothing, while protecting the wearer from high temperatures, may also contribute to heat stress in the wearer. Garments that are acceptable for northern climates may not be safe for hotter southern climates. Heat stress factors must also be considered in selecting personal protective clothing.

Structural Fire Fighting Clothing

Historically, the structural fire-fighting ensemble is the traditional attire worn by firefighters in the fire service. It consists of head and face protection in the form of helmet, protective hood, and eye protection; upper body protection in the form of a turnout or bunker coat; lower body protection in the form of turnout or bunker pants; foot protection in the form of waterproof boots; and gloves for hand protection (Figure 7.18). Originally addressed in the 1975 version of NFPA 1971, (*Standard for Protective Clothing for Structural Fire Fighting*), it only covered the coat and pants. Other standards covered the other portions of the ensemble. In 1997, a new edition of NFPA 1971 was published (*Standard on Protective Ensemble for Structural Fire Fighting*), incorporating all of the various individual standards. The greatest change in the standard occurred in 1991 when third-party certification and testing of the equipment, labeling, and listing of the equipment was mandated. The health and safety officer and anyone responsible for purchasing protective clothing must be familiar with the requirements of this standard.

Inspection

As established in the department's protective clothing policy, inspection of the structural fire fighting

Figure 7.18 Protective clothing is usually stored in open, ventilated cages in or near the apparatus room.

ensemble occurs at the time new clothing is received, periodically by the health and safety officer or unit supervisor, and following its use. These inspections include the following elements:

- *Initial inspection* — The health and safety officer, a designated member of the health and safety committee, or the staff officer in charge of equipment purchases inspects to ensure that the clothing meets the department's specifications. Unsatisfactory or damaged clothing must be returned to the vendor for credit, replacement, or repair.

- *Periodical inspection* — The inspecting officer inspects clothing for proper condition and cleanliness. At this point, the inspecting officer requires that the ensemble be cleaned, repaired, or replaced depending on its condition.

- *Inspection following each use* — The individual firefighter inspects the clothing, cleans it, makes minor repairs, or submits it for major repairs as needed. The department should maintain a supply of spare clothing for immediate replacement if possible.

A standardized criteria for replacement should be included in the protective clothing policy. The need for replacement should be determined by the age and condition of protective clothing. Currently, there is no set replacement standard available. It is recommended that the department establish a replacement standard that allows the entire inventory of protective clothing to be replaced on a specific schedule such as five, seven, or ten years or as needed. Most manufacturers recommend keeping clothing in service no longer than ten years.

Care, Cleaning, and Maintenance

Personal protective clothing used in structural fire fighting usually gets the hardest and dirtiest wear of all the equipment provided. The protective clothing program should include instructions on the care, cleaning, and maintenance of the clothing. The manufacturer's guidelines are one source for this information. Further sources include NFPA 1500, Appendix 5-1; NFPA 1581, *Standard on Fire Department Infection Control Program*; or the *PPE Care and Use Guidelines* published by Southern Area Fire Equipment Research (SAFER) and Fire Industry Equipment Research Organization (FIERO). Current addresses for these last two fire service organizations are found in the appendix of NFPA 1971.

Care of structural fire fighting clothing includes the proper storage of garments both on the apparatus and at the station. Protective clothing can be damaged by physical contact with portions of the apparatus, in compartments, and near loose equipment. PPE clothing is also very susceptible to damage from the sun. Ultraviolet light can deteriorate the fabric and cause damage to the seams of the outer shell and inner liner. Clothing must be stored out of direct sunlight and never dried in direct sunlight. When stored at the station between shifts, the clothing should be in a well-ventilated locker outside the living quarters. This ventilation helps to prevent mildew and mold from growing on the garments and keeps contamination out of the living or dorm areas.

Proximity Clothing

Proximity fire fighting involves operations at incidents such as aircraft fires, bulk flammable liquids fires, and bulk flammable gas fires that involve high levels of conductive, convective, or radiant heat (Figure 7.19). Clothing intended for this type of fire fighting is defined in NFPA 1500 and NFPA 1976, *Standard on Protective Ensemble for Proximity Fire Fighting*. The ensemble includes coat, hood, footwear, gloves, and trousers that meet the requirements of NFPA 1976 (Figure 7.20). NFPA suggests that SCBA be worn internally rather than outside the proximity clothing. This placement protects the SCBA and its fittings from extreme radiant heat.

Because of the specialized design and material used in proximity clothing to reflect high levels of heat, follow the manufacturer's recommendations for care, inspection, and cleaning. Like structural clothing, proximity clothing is required to be labeled and third-party certified. NFPA 1976 also contains care and cleaning recommendations.

The health and safety officer must remember that proximity clothing is not entry clothing and is not intended for protection against prolonged direct flame contact. Ensembles designed for entry into incidents that may involve direct flame con-

Figure 7.19 Proximity suits allow firefighters to make close approaches to extremely hot fires.

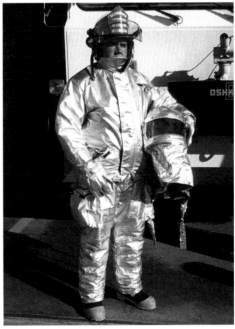

Figure 7.20 A typical proximity suit.

tact are called entry fire fighting ensembles and are defined in NFPA 1976 (Figure 7.21). The limitation of proximity clothing should be emphasized to all personnel during training.

Emergency Medical Services (EMS)

With the increase in the variety and quantity of infectious disease, firefighters have become very concerned with personal protection during emergency medical responses (Figure 7.22). While those personnel who perform both EMS and structural fire fighting duties tend to depend on their structural clothing for protection, personnel assigned to medical units require specialized clothing. Protection against liquid-borne pathogens takes the form of garments, gloves, and face protection that meet the requirements of NFPA 1999. This standard is compatible with the other protective clothing standards in its requirements for testing, labeling, and third-party certification during manufacture. Procedures for cleaning and disinfecting reusable EMS protective clothing are covered in Chapter 5, Facilities Safety.

Although budget constraints may tempt smaller fire departments to depend solely on structural fire fighting clothing for all types of responses, including EMS, they should not allow this to happen. Not only does it place firefighters at risk, but it places victims at risk as well. Fire fighting clothing exposes the victim to contamination and further risk of infection. The health and safety officer must provide the administration with supporting documentation, including NFPA and OSHA requirements, that justify the funding to support proper medical operations.

Hazardous Materials Response Clothing

Hazardous materials (haz mat) operations can take place at manufacturing sites, storage sites, dispensing sites, or transportation locations including highways, airports, railroad lines, or marine facilities. Protection against hazardous materials includes proximity clothing (if the material is involved in fire), vapor protection, splash protection, and support function protection if the material is not on fire. Design, certification, and testing requirements are found in the appropriate NFPA

Figure 7.21 Entry level suits allow the wearer to work in a total flame environment. Such types of garments are extremely expensive and rarely found in municipal fire departments. *Courtesy of Fyrepel, Inc.*

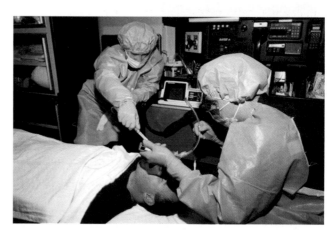

Figure 7.22 EMS personal protective equipment is tested and certified for protection against body fluids. *Courtesy of Mike Nixon.*

standards, including 1976, 1991, 1992, and proposed 1994. Additional information can be found in OSHA Title 29 *CFR* 1910.120 and Appendix A and Appendix B of the IFSTA **Hazardous Materials for First Responders** manual.

Not all fire departments have the resources to purchase the necessary types of hazardous materials response clothing to meet all potential hazards. Proper protective clothing is not the only issue, though. The fire department should be clear on its mandate from the municipality or authority having jurisdiction. If it is the duty of the fire department to respond to hazardous materials incidents, then it must also have the training, funding, and written policy to support this activity. Departments that are not funded or equipped to respond to haz mat incidents should not attempt it. The health and safety officer, working with the hazardous materials officer or the health and safety committee, must determine the recommended policy on what level of haz mat response is provided by the fire department. This determination is done through the risk management model and involves analyzing the potential hazards, determining the estimated frequency of incidents, and determining the most cost-effective response. The department should also consider the cost effectiveness of single-use garments over reusable garments. This consideration includes the possibility that permanent reusable garments may deteriorate with time because they have never been used.

The Environmental Protection Agency (EPA) has developed four levels of protection — Level A, Level B, Level C, and Level D — against hazardous chemical exposure. The EPA defined these levels of protection for personnel assigned to hazardous waste sites where emergency conditions do not normally exist. The levels of protection are described as follows:

- *Level A protection* — Worn when the highest level of respiratory, skin, eye, and mucous membrane protection is needed; consists of positive-pressure self-contained breathing apparatus and fully encapsulating chemical-resistant suit (Figure 7.23).

- *Level B protection* — Worn when the highest level of respiratory protection is needed with a lesser level of skin and eye protection; consists of positive-pressure self-contained breathing apparatus and a chemical-resistant coverall or splash suit. The Level B ensemble is the minimum level recommended for initial-site entries until the health and safety officer defines the hazards by monitoring, sampling, and other reliable methods of analysis (Figure 7.24).

- *Level C protection* — Worn when the type of airborne substance is known, the concentration has been measured, criteria for using air purifying respirators is met, and skin and eye exposure are unlikely; consists of air purifying respirator and chemical-resistant coverall or splash suit. Periodic air monitoring is required.

- *Level D protection* — Worn when respiratory or skin hazards do not exist; primarily a work uniform.

These levels do not mirror the requirements of NFPA 1991 and 1992. Each EPA level does not

Figure 7.23 Chemical protective clothing is worn by personnel who may be in contact with hazardous chemicals. *Courtesy of Joe Marino.*

Figure 7.24 This photo shows Level B suits with SCBA worn on the outside of the garment.

define the chemical protective clothing with respect to the use based upon the risk or hazard present and the required performance of the protective clothing and equipment (Figure 7.25). Selecting protective clothing and equipment solely based on design and configuration is not sufficient to ensure proper protection for the user. EPA levels of protection do not specify what performance (for example, vapor-protective or liquid-splash protection) the selected protective clothing must offer. For firefighters and other emergency response personnel, the only acceptable type of protective clothing includes full or total encapsulating suits or nonencapsulating or splash suits coupled with accessory clothing such as chemical-resistant gloves and boots as defined by the NFPA standards (Figure 7.26). Both NFPA 1991 and 1992 establish a minimum level of limited chemical flash fire protection not found in the EPA requirements.

Figure 7.25 NFPA 1991 vapor protection personal protective equipment worn while placing contaminated material in an overpack drum.

Figure 7.26 Splash protection is provided by NFPA 1992 compliant garments sealed with duct tape.

Wildland PPE

Wildland fire fighting is defined by NFPA 1977, *Standard on Protective Clothing and Equipment for Wildland Fire Fighting,* as "the activities of fire suppression and property conservation in woodlands, forests, grasslands, brush, prairies, and other such vegetation, or any combination of vegetation that is involved in a fire situation but is not within buildings or structures." Personal protective equipment for wildland fire-fighting operations includes protective clothing (shirt, jacket, one-piece garment, and trousers), helmet, gloves, footwear, protective shelter, and protective face/neck shroud (Figure 7.27). One of the factors or issues affecting the wildland firefighter is heat stress. The objective of this protective clothing is to provide thermal protection for the firefighter but not to increase internal heat stress. Therefore, clothing designed to provide the thermal protection for structural fire fighting is not applicable for extended wildland operations.

Figure 7.27 NFPA 1977 compliant equipment supplies full protection for personnel fighting wildland fires. *Courtesy of Monterey County Training Officers.*

NFPA 1977 defines the certification, labeling and information, design requirements, performance requirements, and test methods for each of the components of wildland protective clothing. The protective clothing and equipment inspection and care requirements are specified in NFPA 1977, NFPA 1500, and Title 29 *CFR* 1910.132. Additional requirements, which include carrying water and food supplies as part of the protective ensemble, may be found in the federal and state forestry regulations (Figure 7.28).

Confined Space and Structural Collapse PPE

Rescue operations in confined space and structural collapse situations present a unique set of conditions. The protective clothing and equipment normally worn by firefighters may be too bulky and cumbersome and even unnecessary in these operations. While structural fire fighting clothing may be worn in some instances, tight spaces require specialized personal protection. There is currently no NFPA standard that covers these types of incidents; however, NFPA 1951, *Standard on Protective Ensemble for Urban Technical Rescue Incidents,* is in development. The health and safety officer has to rely on the OSHA requirements found in Title 29 *CFR* 1910.146 (k).

Equipment needed for personal protection in confined space and structural collapse situations consists of hard hat or helmet, goggles, gloves, boots, life-safety harness, and life-safety ropes and accessories and possibly including respiratory protection (Figure 7.29). Station/work uniforms provide limited torso and arm protection. A lightweight jacket can be used to protect against abrasions. All items must meet the requirements of OSHA or the pertinent NFPA standard on life-safety ropes and system components (NFPA 1983). Inspection, care, and maintenance follow the requirements of the equipment manufacturers.

Station Wear/ Work Uniform

Station wear and work uniforms are intended to provide a limited level of safety for the employee in the event of an unexpected exposure to fire (Figure 7.30). These uniforms are not considered a primary level of protection like the structural or proximity ensembles. In addition, the station/work wear uniform should not contribute to injury in the event of exposure to fire. According to NFPA 1975, station/work uniforms shall be constructed of either natural fabrics such as pure wool or cotton or flame-resistant fabrics such as Nomex®, Flamex®, or PBI® and Kevlar®.

Figure 7.28 In addition to the personal protective equipment worn while fighting forest fires, drinking water and other supplies are also carried. *Courtesy of NIFC.*

Figure 7.29 Prior to entering a confined space, the health and safety officer must ensure that all personnel are properly equipped.

The basic elements of the station/work uniform are shirts, trousers, and coveralls. Underwear, socks, dress uniforms, and outerwear jackets are excluded from NFPA 1975 requirements. The health and safety officer should emphasize to fire department members that underwear and socks made of syn-

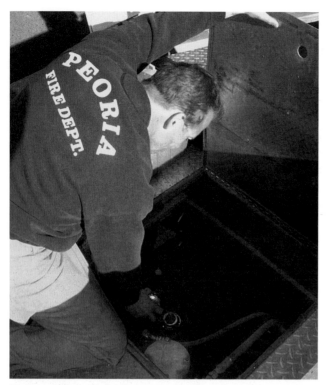

Figure 7.30 Station wear that is worn while inspecting and cleaning apparatus must meet NFPA 1975 standards.

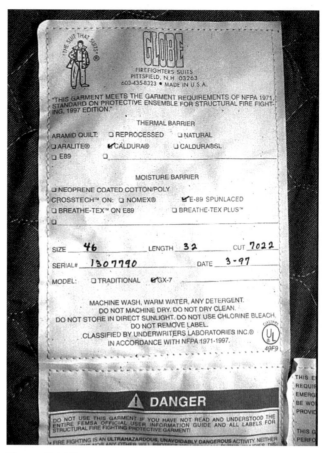

Figure 7.31 Personal protective equipment must be cleaned in accordance with the manufacturer's recommendations that are included on a tag in the garment.

thetic fabrics can contribute to burns if the primary protective garment is severely damaged. Members should be encouraged to select clothing and undergarments that are made from natural fibers.

Because NFPA 1500 requires that fire departments provide for the cleaning of station/work uniforms, the health and safety officer should include cleaning and inspection instructions in the protective clothing program. The program includes requirements for periodic inspections of the clothing to ensure proper cleaning and maintenance. The program also provides for a commercial cleaning service or internal cleaning procedures and a policy on retiring old clothing. The health and safety officer must ensure, through the inspection program, that all station/work uniforms meet the requirements of NFPA 1975. All garments must bear the original manufacturer's label and certification. There are documented occasions where employees have attached old NFPA-compliant labels to noncompliant clothing in an attempt to circumvent the requirements and save money. The health and safety officer or inspecting officer must actively work to prevent this unsafe act.

Cleaning of station/work uniforms must comply with the manufacturer's recommendations. If uni-forms are cleaned in department-provided clothes washers, those washers must be dedicated solely to the cleaning of these items. NFPA 1581 establishes the guidelines for this cleaning (Figure 7.31). Chapter 5, Facilities Safety, contains information on the cleaning and disinfecting of station/work uniforms.

There are no guidelines for how long station/work uniforms should remain in service. If the garments are made of natural fibers, they can remain in use as long as they retain a uniform appearance and have no physical damage due to burns, tears, or rips. Fire-resistant uniforms, however, are another matter. The longevity of garments with chemically added fire-resistant properties is based on the number of washings — usually about 100 times. Washing reduces the fire resistance of some fabrics. Uniforms that have inherent fire-resistant properties that cannot be washed out last longer — usually up to five years. The manufacturer's recommendations should be consulted if there is a question of longevity of a specific garment.

Foot protection is one element of the station/work uniform that is not addressed by NFPA 1975. Requirements for different types of work shoes can be found in the ANSI standard Z41 latest edition *American National Standard for Personal Protection-Protective Footwear* (Class 75 toe protection is recommended). OSHA requirements can be found in Title 29 *CFR* 1910.136. The health and safety officer should enforce these requirements through the protective clothing program. Work shoes can provide protection against rolling or falling objects and/or objects piercing the shoe sole.

 ## Personal Protective Devices

Personal protective devices include a variety of nonclothing type items such as personal alert safety system (PASS) devices, life-safety harnesses, eye protection, and hearing protection.

PASS Device

One of the newest additions to the firefighter's ensemble is the personal alert safety system or PASS device (Figure 7.32). It provides an audible alert tone as well as a visible warning light in the event that the wearer is incapacitated and the wearer needs assistance. The unit activates when it remains motionless for 30 seconds. The units currently take two forms: units that attach to the protective clothing and must be turned on manually and units that are integrated into the SCBA and are turned on when air is flowing through the system. Since the PASS device only provides protection when it is turned on, the advantage of the integrated system that is linked to the operation of the SCBA is obvious. However, the integrated unit is of no value when the SCBA is not in use such as when firefighters are in confined spaces or at wildland operations. The health and safety officer, when advising the department on the selection of PASS devices, must take into consideration the best type of unit for the functions to be performed. In some cases, departments have chosen to supply both types of PASS devices.

According to NFPA 1500, PASS devices must be tested at least weekly and before each use (Figure 7.33). The department must provide training in its use and testing. Units that fail to operate must be

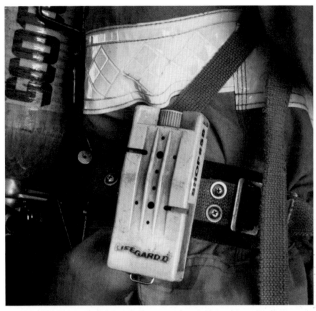

Figure 7.32 The PASS device is worn on the protective clothing, usually on the SCBA harness.

Figure 7.33 The PASS device must be tested weekly and before each use to ensure that it is operating correctly.

replaced immediately. Most units require maintenance by the manufacturer; therefore, the department should maintain spare replacement units. All units should be inventoried, and records should be kept on the devices' assignment, use, testing, and maintenance. Design and testing criteria of PASS devices are found in NFPA 1982, *Standard on Personal Alert Safety Systems (PASS)*.

Life-Safety Harnesses

According to NFPA 1983, life-safety harnesses include harness assemblies that are used to support people during rescue operations (Figure 7.34). These assemblies include devices that are used to support both victims and firefighters during rescue, fire fighting, and training operations. The standard divides life-safety harnesses into three categories:

• *Class I* — Harness that fastens around the waist and around the thighs or under the buttocks and is designed to support the weight of one person on a ladder or emergency escape; designed to support 300 pounds (136 kg) of weight (Figure 7.35).

• *Class II* — Harness that fastens around waist and around thighs or under buttocks and is designed to support the weight of two persons during rescue; designed to support 600 pounds (272 kg) of weight (Figure 7.36).

Figure 7.34 The life-safety harness must be designed to prevent the wearer from becoming inverted while rappelling. *Courtesy of SKEDCO, Inc.*

Figure 7.35 Ladder belts are considered Class I life-safety harnesses and are used when working on the tip of an aerial device.

- *Class III*—Harness that fastens around the waist, around the thighs or under the buttocks, and over the shoulders, designed for two persons during rescue and where inverting is possible; designed to support 600 pounds (272 kg) of weight (Figure 7.37).

Life-safety harnesses must be inventoried (with individual serial numbers), inspected and tested periodically, and stored in a manner that prevents damage to them. Cleaning and repair should conform to the manufacturer's instructions. Life-safety harnesses should be stored in bags to prevent the effects of moisture and sunlight on the material.

As part of his overall record-keeping duties (Figure 7.38), the health and safety officer should maintain a record of all safety-related harness issues including date of purchase, specifications, inspections, testing, repairs, and use.

Figure 7.36 Class II life-safety harnesses must be able to support the weight of two persons.

Figure 7.37 Class III life-safety harnesses are designed to prevent the wearer from becoming inverted during the rope rescue incident.

Figure 7.38 The health and safety officer must ensure that records are maintained on all life-safety harnesses including date of purchase, specs, inspection, testing, repairs, and use.

Eye Protection

Eye and face protection required by NFPA 1500 must be appropriate to the hazard to which the employee is exposed and must meet the requirements found in ANSI Z87.1, *Practice for Occupational and Educational Eye and Face Protection*. This protection includes the following components:

- *SCBA full facepiece* — Designed to protect the respiratory system, eyes, and mucus membranes (Figure 7.39)

- *Helmet-mounted faceshield* — Provides limited protection to the eyes from flying debris (Figure 7.40)

- *Safety glasses* —- Provides limited protection to the eyes from flying debris (Figure 7.41)

- *Helmet-mounted safety goggles* — Provides full protection to the eyes from flying debris and chemical splash (Figure 7.42)

- *Eye protection for welding/cutting* — Provides protection to the eyes from ultraviolet rays and flying debris found in the welding/cutting process (Figure 7.43)

The health and safety officer must select the proper type of protection for the specific task, provide a policy outlining the use of the eye protection, and train employees in the proper use and care of the items.

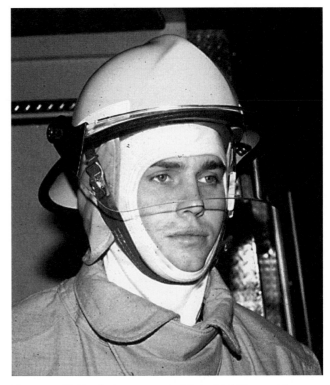

Figure 7.40 The helmet-mounted faceshield is used to provide protection to the eyes from flying debris.

Figure 7.41 Safety glasses privide limited eye protection.

Figure 7.39 Eye protection is provided by the SCBA facepiece when the unit is worn as part of the protective equipment ensemble.

Figure 7.42 Safety goggles provide eye protection from flying debris and chemical splash.

Figure 7.43 Welding/cutting shields provide protection from sparks, debris, and the ultraviolet rays emitted by the welding/cutting process.

Hearing Protection

Firefighters are exposed to harmful levels of noise daily. Sources of noise are the audible warning devices on the apparatus, apparatus engine and pump operations, power-operated tools and equipment (such as radios, lawn mowers, and chain saws), and the emergency scene itself (Figure 7.44). NFPA 1500 requires that hearing protection be provided anytime the ambient noise level exceeds 90 decibel (dB).

The health and safety officer must develop a hearing conservation program for the department. The program must meet the OSHA requirements specified in Title 29 *CFR* 1910.95 whenever noise exposure exceeds or equals 85 dBA for an 8-hour time-weighted average (TWA) (Figure 7.45). The

Figure 7.44 Hearing protection must be worn while operating power tools.

Figure 7.45 The health and safety officer is responsible for performing sound level testing for department facilities, apparatus, and equipment. *Courtesy of Mike Nixon.*

goal of this program is to identify the various sources of noise in the workplace and reduce or eliminate the hazard. As part of the health and wellness program's annual medical evaluations, hearing tests must be given to all members. A record of the results must be retained by the department and provided to the member. Sound-level testing should be performed at all facilities, on all apparatus and equipment, and on all newly purchased equipment. As part of the hearing conservation program, the health and safety officer should attempt to reduce noise to an acceptable level. If this is not possible, hearing protection must be provided for employees, and warning signs must be posted in the effected areas.

Hearing protection takes the form of apparatus-mounted radio communication headsets, earmuffs, or single-use or reusable earplugs. Whichever type is selected, it must meet the standard to protect the individual from noises in excess of 90 dB (Figure 7.46).

 ## Auxiliary Safety Equipment

Auxiliary safety equipment includes life-safety rope and system components. Normally, this equipment is not issued to individual firefighters and may be assigned only to a specialized rescue unit within the department. It may also experience only infrequent use. Lack of use requires that it be inspected regularly, usually on a weekly basis. Record keeping is also an integral part of the proper use and maintenance of auxiliary safety equipment.

Life-Safety Rope

Life-safety rope is used strictly for supporting the weight of firefighters and/or victims during fire fighting, rescue, or training operations. Its use is regulated by NFPA 1500 while its construction and testing are regulated in NFPA 1983. Qualified inspectors must inspect the rope, and documentation on each rope must be maintained (Figure 7.47). The manufacturer provides inspection requirements and inspector qualifications. The health and safety officer must include information on the life-safety rope along with the life-safety harness.

Guidelines for use, storage, cleaning, inspecting, and disposal of life-safety rope must be included in

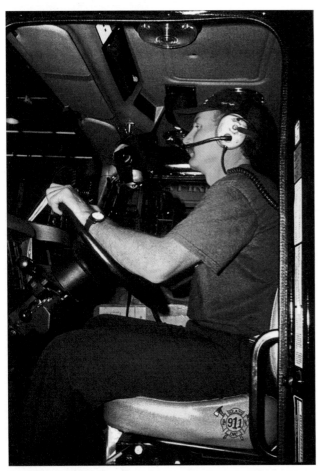

Figure 7.46 Hearing protection can be provided in apparatus with headset communications systems that allow the crew to communicate with each other.

Figure 7.47 Life-safety rope must be inspected following each use.

the department's policy and procedures manual or standard operating procedures manual. Rope that fails the test or has been used for any purpose other than life safety must be removed from service as life safety rope and destroyed or used in another capacity.

Life-Safety Rope System Components

Life-safety rope system components include carabineers, ascent and descent devices, rings, snap-links, pulleys, blocks and tackles, and other devices used with rescue rope (Figure 7.48). NFPA 1983 provides the health and safety officer with the requirements for this equipment. Guidelines for inspection, care, maintenance, use, and testing are the same as those for the life-safety harness and ropes.

 ## Miscellaneous Safety-Related Apparel

Safety-related apparel such as items associated with water rescue and highway operations is controlled by regulatory bodies such as the United States Coast Guard, the Department of Transportation, and OSHA. These items are also of concern to the health and safety officer.

Water Rescue

Because bodies of water are found in or near most jurisdictions that provide fire and rescue services, including the traditionally arid Southwestern United States, departments should be prepared to perform water-related rescues. Through the risk management model, the health and safety officer determines the type of protection that is required; structural fire fighting clothing may not be suitable. Protection includes head protection in the form of a hard hat or helmet, eye protection, gloves, safety harness, and personal flotation device (PFD). The United States Coast Guard regulates the type and style of personal flotation devices used. The U.S. Coast Guard also governs the style of rescue rings and other flotation devices used (Figure 7.49).

Personal flotation devices, like the rest of the safety equipment, must be inspected regularly, maintained, and removed from service at the end

Figure 7.48 The health and safety officer must inspect the life-safety components periodically.

Figure 7.49 Water rescues, especially in colder climates, require specialized personal protective equipment.

of its manufacturer's-recommended life span. All members who participate in water rescue must be thoroughly trained in the use of the rescue equipment.

Highway Operations

Firefighters responding to vehicle incidents, fires, medical emergencies, and hazardous materials operations on the nation's highways are in double jeopardy. First, there is the inherent danger from the incident itself. Second, firefighters become potential targets for vehicles that are passing the incident. Inattentive or reckless drivers can injure or kill emergency responders who are trying to perform their duties. The health and safety officer must establish operational protocol for highway operations that reduces the risk to responders. This protocol includes the proper positioning of apparatus (Figure 7.50) and requiring responders to wear garments with reflective trim (Figure 7.51). Reflective trim is required on all structural protective clothing currently manufactured in the United States and Canada. When this clothing is not in use at highway incidents, lightweight safety vests with reflective trim must be provided to responders to increase their visibility. Safety vests are usually made from synthetic fabrics and, therefore, must not be worn when there is a danger of fire. Additional requirements are found in NFPA 1500 in the section titled Emergency Scene Operations.

Figure 7.50 The health and safety officer should initiate a protocol for parking apparatus to form a protective barrier.

Figure 7.51 Reflective trim on personal protective equipment provides increased visibility for firefighters working at night.

 ## Summary

No matter what the hazard is that the individual firefighter must face, he must be equipped with the correct personal protective apparel and equipment. It is the duty of the health and safety officer to identify the safety hazards; specify the appropriate level of protection; develop a policy regarding the use, care, cleaning, maintenance, inspection, and testing of the items; and ensure that training is provided. All of these provisions can be accomplished through the creation and administration of the NFPA 1500 mandated protective clothing program.

Emergency Response and Apparatus/Vehicle Safety

8

This chapter provides information that will assist the reader in meeting the following job performance requirements from NFPA 1521, *Standard for Fire Department Safety Officer,* 1997 edition.

3-3.1 The health and safety officer shall ensure that training in safety procedures relating to all fire department operations and functions is provided to fire department members. Training shall address recommendations arising from the investigation of accidents, injuries, occupational deaths, illnesses, and exposures and the observation of incident scene activities.

3-7.1 The health and safety officer shall review specifications for new apparatus, equipment, protective clothing, and protective equipment for compliance with the applicable safety standards, including the provisions of Chapters 4 and 5 of NFPA 1500, *Standard on Fire Department Occupational Safety and Health Program.*

3-7.2 The health and safety officer shall assist and make recommendations regarding the evaluation of new equipment and its acceptance or approval by the fire department in accordance with the applicable provisions of Chapter 4 of NFPA 1500, *Standard on Fire Department Occupational Safety and Health Program.*

3-7.3 The health and safety officer shall assist and make recommendations regarding the service testing of apparatus and equipment to determine its suitability for continued service and in accordance with Chapter 4 of NFPA 1500, *Standard on Fire Department Occupational Safety and Health Program.*

Chapter 8

Emergency Response and Apparatus/Vehicle Safety

Although most people think of fire engines, or pumpers, when they think of fire department apparatus, pumpers make up only a portion of the modern fire department fleet. In addition to pumpers, fire departments may own and operate aerial ladders or elevating platforms, wildland fire apparatus, tankers/tenders, ambulances, heavy/light rescue trucks, hazardous materials trucks, mobile command centers, aircraft crash/fire-rescue vehicles, fuel and supply trucks, staff cars, and battalion/district chief's vehicles (Figure 8.1). It is the duty of the health and safety officer to ensure that this equipment meets the needs for which it was purchased and the safety standards of the department. He is also responsible for ensuring that personnel are trained in the safe operation and care of the apparatus/vehicles. To assist with the realization of these goals, the safety officer must perform the following activities:

- Review vehicle specifications during the design and purchasing phase in order to make recommendations for safety-related changes.

- Witness acceptance testing and periodic testing of the apparatus to ensure that it meets the applicable NFPA standards.

- Ensure safe operating practices.

- Investigate vehicle incidents that result in damage to apparatus, vehicles, and property.

- Correct unsafe practices.

- Provide guidelines for driver/operator training and certification programs.

Even though the trend in fire apparatus design during the last quarter of the twentieth century has

Figure 8.1 Fire departments operate a variety of apparatus as illustrated in this photograph. *Courtesy of Chris Mickal.*

been toward increased safety, firefighter injuries and fatalities related to vehicle operation continue to occur. NFPA reported that 26 percent of all firefighter fatalities between 1987 and 1996 resulted from vehicle/apparatus-related incidents. Three percent of fatalities were the result of falls from apparatus. By 1999, the percentage of vehicle/apparatus-related fatalities was 28 percent or 32 of the 112 total deaths. It is the duty of the health and safety officer to prevent these types of injuries and fatalities through apparatus design standards, driver/operator training, and apparatus/vehicle safety policies.

 Apparatus and Vehicle Safety Action Plan

The fire department should evaluate all apparatus/vehicles currently in use and establish an action plan to address and correct any existing safety issues or deficiencies. The health and safety officer should take a leading role to ensure that this action plan is developed and implemented (Figure 8.2).

The apparatus and vehicle safety action plan of the fire department includes a wide spectrum of issues such as the following:

• Specifications

• Design

• Construction

• Purchase

• Operation

• Response procedures

• Inspection

• Maintenance

• Repair

The health and safety officer must also ensure that the manufacturer is licensed to sell and service

in his state. Purchasing from a state or province licensed dealer ensures that the dealer is legally bound to comply with warranty agreements in the contract.

Design and Review of Fire Apparatus Specifications

The health and safety officer plays an integral part in the design and review of new fire apparatus specifications. During the initial specification phase, the following elements must be addressed to ensure personnel safety:

• *Visibility* — Apparatus visibility is an important aspect in developing specifications. The health and safety officer must work with the apparatus supervisor or committee to ensure that the requirements of the appropriate NFPA standard are met. These requirements include adequate warning lights, reflective trim, and an appropriate color scheme for the apparatus (Figure 8.3).

• *Audible warning devices* — The health and safety officer must stay current with developments in

Figure 8.3 Apparatus visibility can be increased by the vehicle color scheme, placement of warning lights, and the use of reflective trim.

Figure 8.2 The health and safety officer, working closely with the apparatus chief, should be involved in the apparatus design process.

the field of warning devices and also with changes in the NFPA apparatus standard. Audible warning devices, including sirens, air horns, and backup alarms, must meet the NFPA standard while not creating a safety problem for the crew (Figure 8.4). The location of the audible devices must provide the greatest sound coverage to the front of the apparatus. At the same time, the crew must not be subjected to noise levels greater than 85 dB. For this reason, audible warning devices may not be mounted on the roof of the crew compartment. The health and safety officer also must ensure that the apparatus electrical system is capable of handling the added load. This electrical capability is also an issue when upgrading the warning devices on existing apparatus.

Additional areas of concern that should be addressed during the design phase include the following:

- Obstructions to the driver/operator's line of sight such as window posts, spotlights, or other crew members (Figure 8.5)

Figure 8.4 Some apparatus are equipped with air horns.

Figure 8.5 The driver/operator's view may be obstructed by the cab corner post and other items mounted on the dashboard.

- Proper placement of handrails or grab rails in the cab and on the exterior of the apparatus
- Proper placement of access steps into the cab, hose bed, turntable, or upper body
- Ergonomically correct storage access (Figure 8.6)
- Types of terrain in which the apparatus operates
- Department of Transportation (DOT) and state/ provincial restrictions on gross vehicle weight and axle/tire loads

During the design and specification phases, the department may be confronted with the choice of purchasing a custom-built apparatus or an apparatus mounted on a commercial chassis. The health and safety officer consults with the apparatus chief or committee to determine the safest, most economic selection. Maintenance records, apparatus incident investigations, and surveys of designs from other cities can provide the information needed to make an informed decision.

Applicable Standards

The health and safety officer must be familiar with the apparatus requirements found in NFPA 1500, 1901, and 1906. NFPA 1901, *Standard for Automotive Fire Apparatus*, covers the design and construction of pumpers, initial fire attack vehicles, tankers, ladders and elevating platforms, and special service apparatus. NFPA 1906, *Standard for Wildland Fire Apparatus,* provides the minimum requirements for apparatus primarily designed and deployed to fight wildland fires. Appendices in both NFPA 1901 and 1906 contain helpful suggestions for writing specifications.

Figure 8.6 The health and safety officer must ensure that storage of equipment does not create a safety hazard for firefighters.

NFPA 1901 Requirements

NFPA 1901 provides the minimum requirements for the following apparatus:

- Pumper fire apparatus
- Initial attack fire apparatus
- Mobile water supply fire apparatus
- Aerial fire apparatus
- Quint fire apparatus
- Special service fire apparatus
- Mobile foam fire apparatus

The standard outlines the basic components of each type of apparatus. Apparatus components include (but are not limited to) the following:

- General requirements: personnel protection, controls, instructions, stability, performance, roadability, serviceability, and predelivery testing
- Fire pump
- Aerial device
- Water tank
- Hose capacity and storage
- Equipment and tool storage
- Foam proportioning system and foam tank
- Chassis, power plant, and electrical systems

This standard defines all mandatory components that ensure that the fire apparatus meets applicable federal, state, and provincial motor vehicle laws and codes once construction is complete (Figure 8.7).

Appendix B of NFPA 1901 guides the purchaser in providing the manufacturer with pertinent information on apparatus needs. This information enables the manufacturer to prepare a bid and a complete description of the apparatus it proposes to construct (Figure 8.8).

NFPA 1906 Requirements

Wildland fire apparatus consists of a vehicle with a pump, water tank, limited hose capacity, and equipment. Apparatus covered in NFPA 1906 have a pump size ranging from 20 gpm to 250 gpm (76 to 946 L/min) and a minimum water tank capacity of 125 gallons (473 L). Units with larger pumping capacity are covered in NFPA 1901. Fire apparatus designed to NFPA 1906 requirements include built-to-specification apparatus or pump/tank packages designed to slip onto a commercial vehicle chassis or into the beds of pickup trucks. Because of the popularity of Class A foam, the standard also addresses compressed air foam systems (CAFS) (Figure 8.9).

Figure 8.8 Commercial fire apparatus are composed of a commercially available cab, engine, driveline, chassis, custom-built body, fire pump, hose bed, and water tank. Commercial fire apparatus must meet the requirements of NFPA 1901.

Figure 8.7 A single manufacturer usually builds custom fire apparatus with few commercial components. Components must meet all federal, state, and local vehicle laws.

Figure 8.9 Apparatus designed specifically for wildland fire fighting are capable of off-road operation in rugged terrain.

Because wildland fire fighting apparatus must be able to operate on both hard surface roads and rugged terrain, the standard addresses both roadability and performance. Wildland fire apparatus must be able to operate at elevations of 2,000 feet (610 m) above sea level. If the apparatus is to be used regularly at higher elevations, the manufacturer must provide a power plant that can deliver the proper performance level. The apparatus must perform in temperatures ranging from 33°F (0.55°C) to 110°F (43.3°C) and operate on grades of 25 percent. Wildland fire apparatus must meet all requirements of the standard while stationary on a grade of 20 percent in any direction. Gross vehicle weight is another important consideration in the design of wildland fire fighting apparatus. The chassis must be designed to carry the load for which it is intended. In addition, the design must take into consideration the off-road environment in which the vehicle operates. If the vehicle is too heavy or has poor weight distribution, it can become stuck in soft terrain or be subject to overturning.

Emergency Operations of Apparatus

State/provincial laws, city ordinances, and departmental standard operating procedures (SOPs) regulate the emergency operation of fire apparatus. Members must be aware of all pertinent laws and SOPs. Failure to know or follow department SOPs can have deadly consequences during emergency or even routine situations.

Most laws and statutes concerning motor vehicle operation are legislated at the state or provincial level. Individual states or provinces may define what constitutes emergency vehicles and exempt them from certain laws or statutes. When emergency vehicles are exempt from certain laws or statutes, the driver/operators are allowed to do the following contrary actions:

- Exceed posted speed limits while having due regard for the safety of persons and property and maintaining full control of the apparatus.

- Proceed past any steady or flashing red signal, traffic light, stop sign, or other device indicating moving traffic after coming to a stop and gaining control of the intersection while having due regard for the safety of persons and property.

- Park or stop on roadways while in the performance of job duties.

- Disobey posted regulations governing direction of movement of vehicles and turning of vehicles in specified directions as long as the operator does not endanger life or property.

- Pass or overtake another vehicle at an intersection with due regard to the safety of persons and property.

The fire department should enact specific rules and regulations pertaining to the emergency response of fire department vehicles/apparatus that meet or exceed municipal, state, provincial, territorial, and/or jurisdictional requirements. NFPA 1500 provides specific guidelines for apparatus operations that should be considered when developing these rules and regulations (Figure 8.10).

◆ Ergonomics

As learned in Chapter 5, Facilities Safety, ergonomics is the science of designing tools, equipment, or workstations to accommodate the worker to the tasks performed in order to minimize job-related injuries. The ergonomic process reviews occupational tasks, especially those that are repetitive and those in which numerous occupational injuries have occurred regardless of the severity. Recommendations are then made for changes in the task, the choice of tools, or the design of the tools and equipment.

It is the responsibility of the health and safety officer to understand ergonomics and apply it to the design of fire apparatus and equipment. As tools and equipment are placed and stored on apparatus, the ergonomics issue becomes how accessible are these tools and equipment and how many firefighters are needed to remove them. Apparatus ergonomic conditions of concern to the fire service include the following:

- Height of hose bed

- Location of preconnected hoselines

- Height of crew compartment steps

- Placement of equipment in compartments

- Tool, ladder, and equipment storage (Figure 8.11)

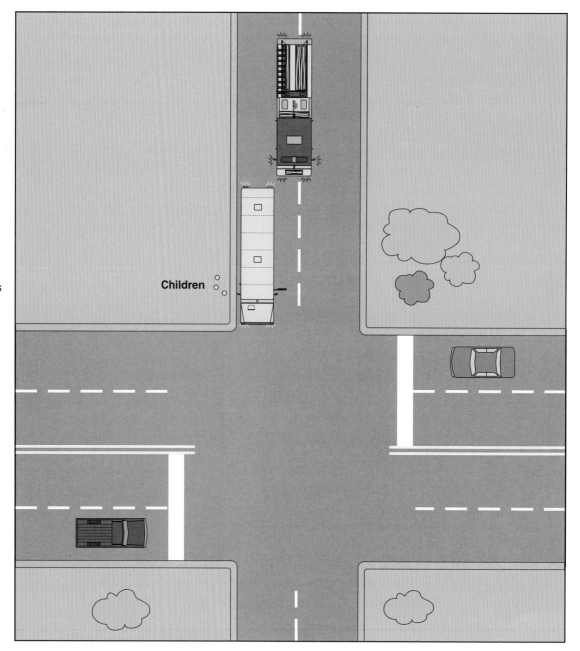

Figure 8.10
Driver/operators must stop their apparatus while school buses are loading or discharging passengers.

Children

Figure 8.11 Firefighters must be able to remove equipment such as ground ladders without injuring themselves.

• Engine and drivetrain access as it applies to both maintenance mechanics and company personnel

◆ Periodic Service Testing

Service testing of fire apparatus is defined in NFPA 1500 as "the regular, periodic inspection and testing of apparatus and equipment, according to an established schedule and guideline, to ensure that they are in a safe and functional operating condition." The health and safety officer should include the requirements for the periodic service testing in the occupational safety program of the department.

NFPA 1500 requires the following service testing for fire department apparatus:

- Fire pumps on all apparatus are to be tested to the requirements of NFPA 1911, *Standard for Service Tests of Fire Pump Systems on Fire Apparatus.* (Figure 8.12)

- Aerial devices are to be inspected and tested in accordance with the requirements of NFPA 1914, *Standard for Testing Fire Department Aerial Devices.*

NFPA 1911 Requirements

According to NFPA 1911, a pumper should be given a service test at least once a year or whenever it has undergone extensive pump or power train repair. These service tests are necessary to ensure that the pumper performs as it should and to check for defects that otherwise might go unnoticed until too late. These service tests include the following:

- Engine speed check
- Vacuum test
- Pumping test
- Pressure control test
- Gauge and flowmeter test
- Tank-to-pump flow rate test

NFPA 1914 Requirements

Service testing for an aerial device ensures that the device offers a minimum degree of safety under continued use. The test requirements in NFPA 1914 specify the frequency of the tests and the procedures to be used. Annual tests consist of visual inspections,

operational tests, and load tests (Figure 8.13). These tests are also conducted following the use of the aerial device under unusual conditions; exposure to excessive heat, stresses, or loads; or any use that exceeds the manufacturer's recommended operating procedures. Complete inspection and testing of the aerial device, including nondestructive testing, must be conducted at least every five years. Nondestructive testing (NDT) must also be conducted when other service tests indicate there is a potential problem with the aerial device (Figure 8.14). Detailed service test information is

Figure 8.13
The annual inspection of aerial apparatus includes the operation of all safety devices.

Figure 8.12 All fire pumps must be tested annually to ensure that they meet the standards of NFPA 1911.

Figure 8.14
Magnetic particle inspection is a form of nondestructive testing and helps find cracks in the aerial device. *Courtesy of Underwriters Laboratories, Inc.*

contained in the NFPA standards and in IFSTA's **Pumping Apparatus Driver/Operator Handbook** and **Aerial Apparatus Driver/Operator Handbook**.

Manufacturers' Service Testing

Each manufacturer establishes criteria for service testing based upon the applicable NFPA testing requirements for pumpers, aerial devices, and special service apparatus. Apparatus must meet the testing criteria before delivery to the fire department. A third-party testing organization such as Factory Mutual or Underwriters Laboratories usually provide personnel for the manufacturer's final certification.

Qualified personnel from the department should also perform an acceptance test on the apparatus when it is received and before it is placed in service. As part of the maintenance program, the manufacturer outlines the schedule for preventive maintenance that fire department mechanics need to follow (Figure 8.15). This maintenance includes annual service testing of pumps and aerial devices (Figure 8.16). Nondestructive testing of aerial devices is required every five years by the NFPA. A qualified third-party testing organization must be hired for the nondestructive aerial device testing.

Refurbished or rebuilt apparatus must also meet the same acceptance criteria and service testing as new apparatus. When an apparatus is refurbished or rebuilt, the health and safety officer must ensure that the result of the work does not compromise safety concerns. In particular, the finished vehicle/apparatus must not exceed the gross vehicle weight that the chassis was originally designed to handle. If the weight exceeds the original design weight, a new suspension is required to compensate for the increased weight. Acceptance testing should focus not only on the refurbished apparatus meeting the design criteria but also on the ability of the power plant, brake system, and suspension to handle the new load.

Insurance Service Office

For departments that are evaluated by the Insurance Service Office (ISO) to determine fire protection insurance rates, proper service testing (at least annually) has to be conducted and prop-

Figure 8.15 Daily apparatus inspections include a check of all fluid levels.

Figure 8.16 Certified apparatus mechanics should check the tightness of turntable bolts with a torque wrench as part of the annual aerial inspection and testing.

erly documented. ISO reviews pump test records for compliance with the requirements of pumping tests in NFPA 1911. In addition, the overall condition, maintenance procedures, and operational procedures of fire apparatus and equipment are evaluated and graded.

Testing and Documentation

Chapter 2 of NFPA 1500 requires that the fire department provide and maintain records relating to the inspection, maintenance, repair, and service of all vehicles used for emergency operations. Personnel who actually perform emergency and scheduled maintenance and inspections on fire department apparatus/vehicles should document these activities. This documentation can also be used for amortization of apparatus/vehicles and for cost analysis for budget preparation.

Testing files that are maintained on each pumper must include the manufacturer's certification for the pump, the acceptance test results, and the annual pump test records. Testing files maintained on each aerial device must include the manufacturer's certification, annual inspection reports, and the third-party certification generated during the five-year test.

The apparatus testing schedule can conform to the preventive maintenance schedule. This reduces the out-of-service time for the apparatus because both testing and repairs can be performed at the same time.

Apparatus Maintenance/ Inspection Program

To ensure the safety of firefighters and to protect the department's financial investment, proper apparatus/vehicle care, inspection, and maintenance are crucial. Therefore, a preventive maintenance/inspection program has to be established to ensure that the necessary maintenance, inspections, and repairs are performed (Figure 8.17). A reliable starting point for a vehicle/apparatus maintenance program is the manufacturer's maintenance instructions or guidelines.

Manufacturers assist with the development of an apparatus/vehicle maintenance/inspection program by establishing specific daily, weekly, monthly, and annual maintenance requirements. Inspection requirements are usually based on industry standards developed in the trucking industry in addition to requirements that are specific to the fire service. For instance, the power plant oil level is checked daily and replaced on a basis of mileage or a set period of time. Brake pads may be inspected

every six months and replaced as needed. Manufacturers of pump and aerial devices establish the schedules for inspections and periodic maintenance for these components.

Apparatus Inspections/Maintenance

The department's standard operating procedures manual must include guidelines for daily apparatus inspection and maintenance. Company personnel must be trained in performing daily inspections and routine maintenance of the apparatus in accordance with these guidelines.

As a minimum during apparatus inspections, fire department personnel should inspect the following parts of the apparatus:

- Tires
- Brakes
- Backup lights and alarm
- Warning lights and devices
- Headlights and clearance lights
- Windshield wipers
- Mirrors
- Seat belts
- Fluid levels (for example, oil, transmission fluid, and fuel)
- Any other item identified by the fire department

The SOP should also contain specific guidelines for inspecting, testing, and maintaining unmanned apparatus stored at remote satellite facilities or kept in reserve.

Figure 8.17 Apparatus may be maintained in a department-operated shop or by a contract vendor.

To prevent an unsafe apparatus from being placed into service, the fire department must develop a list of major mechanical deficiencies. This list is used during each inspection to determine whether the apparatus is safe to operate. If one of these deficiencies is found during an inspection, the fire department apparatus is placed out of service until the repair is made. Items found on a major deficiencies list include the following:

- Brake failure

- Hydraulics system failure

- Failure of warning devices

- Windshield wiper failure

- Failure of headlights, taillights, and directional indicators

- Missing or improperly working seat belts

- Any other condition that would jeopardize the safety of firefighters and the public

All apparatus must be inspected following maintenance and repairs. This inspection is particularly important when repairs have been performed on aerial devices. If the repairs to the aerial device are major, third-party inspection and certification are required.

Apparatus Maintenance Record

Along with the apparatus testing documentation, an apparatus maintenance record must be maintained. This record is the maintenance and repair history of each vehicle/apparatus in the fleet. This type of record can provide justification for apparatus replacement, provide a cost comparison between two similar types of apparatus, establish an amortization or depreciation schedule, and provide a cost analysis for budget preparation.

Qualifications of Mechanics

Only qualified persons — in accordance with the manufacturer's instructions — may perform maintenance, inspection, and repair of fire department apparatus/vehicles. According to NFPA 1500, a qualified person is "a person who, by possession of a recognized degree, certificate, professional standing, or skill, and who, by knowledge, training, and experience, has demonstrated the ability to deal with problems related to the subject matter, the

Figure 8.18 The health and safety officer should assist the apparatus mechanics in obtaining EVT certification.

work, or project." Emergency vehicle maintenance certification programs are available through the National Association of Emergency Vehicle Technicians (NAEVT) (Figure 8.18). The mechanic certification testing is specific to pumps, aerial devices, drive trains, hydraulic systems, and supervisory management skills. Basic qualifications for the certification include the Automotive Service Excellence Certification and NFPA 1071, *Standard for Emergency Vehicle Technician Professional Qualifications,* 2000 edition.

Fire departments must make a commitment ensuring the qualifications of personnel who maintain the department's apparatus/vehicles and to provide continuing education programs for them. Product recalls, technical service bulletins, and part ordering information must be available to the qualified personnel assigned to the vehicle/apparatus maintenance programs. Improper maintenance or lack of maintenance on apparatus and equipment can place a serious liability on a fire department, including jeopardizing the safety and health of firefighters.

Annual Vehicle/Apparatus Inspection

Some states and provinces have requirements that mandate all fire department vehicles/apparatus to be inspected annually. The health and safety officer must research the requirements of his own jurisdiction to determine the requirements that the department must meet.

 ## Specialty Vehicles

A fire department operates a variety of vehicles other than pumpers and aerial devices. Requirements for most of these are found in NFPA 1901 under the classification of Special Service Fire Apparatus. These vehicles can include (but are not limited to) the following:

- Ambulances
- Special rescue apparatus
- Tankers/tenders
- Air supply/light units
- Canteen/rehab units
- Command vehicles
- Trailers
- Wildland or brush trucks
- Aircraft rescue and fire fighting apparatus
- Muster/parade vehicles

Ambulances

Ambulance design is not covered in NFPA 1901. Instead, the design and construction of ambulances must meet an appropriate design criteria such as the *United States General Services Administration (GSA) Federal Specifications for the Star-of-Life Ambulance* or *KKK-A-1822D* (current edition) *Specifications*. This design criteria ensures that the steering, braking, seating capacity, patient care compartment, warning lights, electrical system, emergency medical care equipment, infection control, and other components are incorporated into the construction of the vehicle. The fire department must ensure that the vehicle is inspected before use and that regular preventive maintenance is performed as per the manufacturer's guidelines (Figure 8.19).

Special Rescue Apparatus

This category includes both heavy- and light-rescue apparatus. These types of vehicles can be built on commercial chassis or as custom-designed units. The appendix of NFPA 1901 includes a comprehensive list of the types of tools and equipment normally found on this type of truck. Hazardous materials response apparatus may also be considered as part of this category (Figure 8.20).

Figure 8.19 Equipment storage on similar types of apparatus should be consistent to allow personnel to quickly find the proper items. *Courtesy of Joel Woods, MFRI.*

Figure 8.20 Special rescue apparatus are equipped with specialized rescue tools such as power extrication equipment and lifting bags.

Tankers/Tenders

Adherence to the criteria found in NFPA 1901 ensures that weight distribution, braking capacity, steering, tank baffling, adequate seating capacity, and other crucial components are incorporated into the construction of the vehicle. Fire departments must be very careful when considering the conversion of a commercial vehicle into a fire department tanker/tender. The conversion must be designed to address the identified safety issues to prevent potential hazards that lead to accidents and rollovers. If the vehicle cannot be safely converted, the fire department must not attempt to make this type of conversion (Figure 8.21).

Air Supply/Light Units

These vehicles may have the sole function of supplying breathable air or scene lighting at an incident. Combination units may also be constructed to meet multiple functions. The placement of the air compressor, with regard to noise and vehicle exhaust, must be incorporated into the design of the vehicle. The manufacturer must be consulted to ensure that the weight is properly distributed, braking capacity is sufficient, and the vehicle design is capable of the tasks or operations for which it will be used. Breathing air cylinders must meet federal DOT regulations and American Society of Mechanical Engineers (ASME) specifications. In addition, fragmentation sleeves must be provided in the filling compartment of the apparatus per the current edition of NFPA 1901 (Figure 8.22).

Canteen/Rehab Units

Many fire departments that use canteen/rehab units have purchased and converted a commercial ve-

hicle to serve in this capacity (Figure 8.23). These vehicles must be designed for nonemergency use and must be certified per state motor vehicle codes and inspection requirements. As with any other fire department vehicle, common safety issues focus

Figure 8.22 Mobile air supply trucks provide unlimited breathing air supplies at major incidents.

Figure 8.21 Many departments operate tanker/tenders to ensure large quantities of water where municipal systems are inadequate.

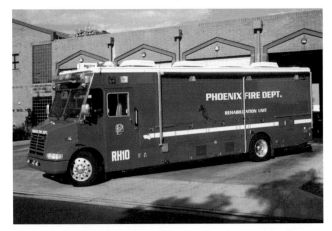

Figure 8.23 Rehabilitation units vary in size and design. Most provide a place for firefighters to rest, cool down, and replenish liquids.

on gross vehicle weight and braking capacity, especially if this vehicle is a conversion unit. Other considerations must include the adequate electrical and power distribution system, adequate seating, noise attenuation, sufficient heating and cooling within the unit, and hygiene facilities.

Command Vehicles

Fire departments either design and construct their own command vehicles from commercial vehicles or have a specialty company construct a command package. Regardless, the fire department must consider the weight and center of gravity that are added to the vehicle chassis and the impact this can have on the braking capacity and cornering of the vehicle. Additional electrical devices such as radios, telephones, laptop computers, warning lights, and other lighting can create overloading problems for the vehicle's electrical system. The fire department must consider all components that will be added to the vehicle and what the impact will be on the vehicle's operation (Figure 8.24).

Trailers

Trailers provide a variety of services to the fire department. Functions include hauling the following equipment:

- Water monitors (Figure 8.25)
- Tractor/trailer units (Figure 8.26)
- Waterborne craft (Figure 8.27)
- Rescue tools and equipment
- Air cylinders
- Potable water supplies
- Generators
- Large diameter hose

Design consideration must be given to the carrying capacity of the trailer, the towing capacity of the powered vehicle, and the types of responses the unit makes. When operating a fire department vehicle that is towing a trailer, safety is paramount. The driver/operator must be properly certified to

Figure 8.24 Command units respond to large-scale operations that involve many agencies or jurisdictions and may last a long period of time.

Figure 8.26 Tractor/trailer units may be used to transport supplies, support mass casualty incidents, or respond to hazardous materials incidents. *Courtesy of Ron Jeffers.*

Figure 8.25 Specialized trailers include large capacity pumping units and hose manifolds.

Figure 8.27 Rescue boats may be towed on trailers by light trucks or pumper apparatus.

drive and operate this type of two-vehicle/trailer combination. If this unit is used for emergency response, a speed limit must be determined based on the size and weight of the combined tow vehicle and trailer regardless of the posted speed limit.

Wildland or Brush Trucks

Apparatus used for wildland operations may consist of a pump/tank package mounted on a pickup or utility vehicle or an all-purpose built, off-road minipumper. Some of the safety issues to be considered are weight capacity, braking capacity, adequate seating, stability (especially off-road), and suspension. Riding on exposed positions on apparatus is not allowed by NFPA 1500 and 1906, and specifically written into the National Wildfire Coordinating Group (NWCG) requirements adopted by the United States Forest Service. Members that operate brush or wildland vehicles must be trained in off-road operations, especially four-wheel drive operations. NFPA 1906 established the criteria for this type of vehicle (Figure 8.28).

Aircraft Rescue and Fire Fighting Apparatus

The NFPA does not provide requirements for aircraft fire fighting and rescue apparatus. Instead, the Federal Aviation Administration (FAA) mandates the specifications for apparatus used for aircraft rescue and fire fighting. These apparatus have special functions that include both hard-surface and off-road operations. Due to the size and weight of these apparatus, driver/operators must have

the proper training to drive and operate them. Because of their size, a police escort may be required if these apparatus must operate on a public street (Figure 8.29).

Muster/Parade Vehicles

These vehicles are strictly for musters, parades, public information events, and nonemergency community activities and are not covered by NFPA criteria. These vehicles may be vintage fire apparatus that have limited seating, yet be a functional piece of apparatus. The fire department must ensure that these vehicles comply with the applicable motor vehicle codes and that they are street worthy. If the apparatus is designed to carry passengers, procedures must outline the safety requirements to use while it is in motion, including the use of safety gates and seat belts (Figure 8.30).

Figure 8.29 Airport fire apparatus are large, heavy oversized vehicles that are designed for a single purpose.

Figure 8.30 Many departments own antique fire apparatus for use in ceremonies and parades. *Courtesy of Joel Woods, Maryland Fire and Rescue Institute.*

Figure 8.28 Apparatus designed for off-road operations may be built on commercial all-wheel drive chassis.

Other Types of Departmental Vehicles

Support and staff vehicles make up a large portion of the fire department's inventory. The majority of these vehicles are commercially built sedans, carryalls, minivans, sports utility vehicles, pickup trucks, flatbed and cargo trucks, and fuel trucks. All will meet federal DOT requirements as constructed. However, if any are to be used for emergency response such as a chief's car, they must meet the same requirements for warning devices and reflective trim (Figure 8.31). The health and safety officer must review the intended use of the vehicles and make suggestions on secure equipment storage, cages to prevent loose equipment from striking personnel, warning devices, and other safety features (Figure 8.32). Departmental policy must address the circumstances for emergency response in these types of vehicles. Vehicles used for field maintenance usually do not respond in an emergency mode and are not equipped with warning devices. These vehicles should be equipped with emergency lighting to keep the vehicle from being struck when parked at the emergency scene (Figure 8.33).

The fuel truck poses the greatest challenge for the health and safety officer. He must ensure that the operator meets the motor vehicle department licensing requirements. These requirements include having a commercial driver's license and passing tests in operation of fuel-hauling vehicles, hazardous materials handling, and the use of fire extinguishers. The only person who may operate this type of vehicle is the qualified operator (Figure 8.34).

All support vehicles should be outfitted with fire extinguishers and first-aid kits. Driver/operators must be trained in the use of these items. They must

Figure 8.31 Vehicles assigned to district or battalion chiefs are usually equipped with audible and visual warning devices. *Courtesy of Joel Woods, Maryland Fire and Rescue Institute.*

Figure 8.32 Staff vehicles, driven by training officers, inspectors, and administrative personnel, are usually marked commercial-type vehicles that lack any warning devices.

Figure 8.33 Apparatus maintenance vehicles provide field repairs that reduce company out-of-service time. *Courtesy of Ron Jeffers.*

Figure 8.34 Most large fire departments operate their own fuel trucks to keep apparatus fueled constantly.

also understand the policies regarding the operation of their vehicles. These policies include the process for reporting damage and injuries resulting from vehicle incidents.

 ## Ventilating, Decontaminating, and Disinfecting Apparatus/ Vehicles

EMS compartments on fire apparatus and other vehicles should be designed to facilitate easy decontamination and disinfecting in the event contaminated materials are placed in these compartments. When designing new apparatus/ vehicles, EMS compartments on fire apparatus, or other rescue vehicles, the fire department must consider engineering controls that aid proper ventilating, decontaminating, and disinfecting procedures (Figure 8.35). Engineering controls are

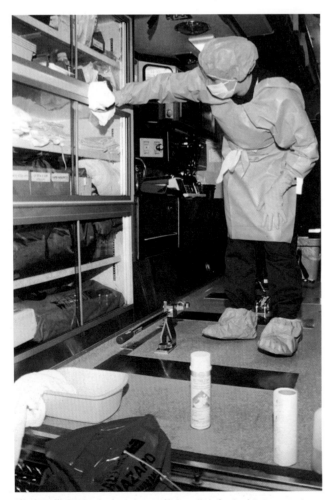

Figure 8.35 Ambulances must be disinfected to prevent the spread of infection. *Courtesy of Mike Nixon.*

built into the vehicle during the design and construction phase. Features include nonporous surfaces both inside and out, filtered ventilation systems, and nonabsorbent washable coverings for seats and cots.

The design criteria for an ambulance or other rescue vehicle that transports patients must take these engineering controls into consideration. The ventilation system must provide complete ambient air exchange in both the driver and patient compartments on a regular basis. Ventilation systems should be controlled in both compartments. Fresh air intakes should be located toward the front of the vehicle to afford maximum intake of fresh air. Exhaust vents should be located in the upper rear of the vehicle. To reduce the risk of tuberculosis infection, high efficiency particle air (HEPA) filters should be integrated into the patient compartment ventilation system.

All seats, mounted cushions, cots, floors, counters, shelves, bulkheads, and container linings must be made of or covered by nonabsorbent, washable materials. These surface materials should be inert to detergents, solutions, and solvents for disinfecting and cleaning as described in the United States Fire Administration's (USFA) *Guide to Developing and Managing an Emergency Service Infection Control Program* (Publication # FA112).

All disinfectants used should be approved and registered with the Environmental Protection Agency (EPA) as a tuberculocidal disinfectant. Members using disinfectants must be aware of safety and health precautions such as ventilation, use of appropriate personal protective equipment, and flammability and reactivity of the disinfectants. Chlorine bleach diluted with water can be used to disinfect compartments, hard surfaces, or other areas of a vehicle or apparatus. An MSDS must also be provided for all types of cleaning solutions used on the vehicle/apparatus.

 ## Apparatus Driver/Operator

The fire apparatus driver/operator is responsible for safely transporting firefighters, the apparatus/ vehicle, and equipment to and from the scene of an emergency or other service call. When the driver/ operator is under the direct supervision of an officer, the officer is accountable for the actions of the driver/operator. Driver/operator guidelines are found in NFPA 1500 and NFPA 1002, *Standard on Fire Apparatus Driver/Operator Professional Qualifications.*

Licensing

In the United States, the Department of Transportation establishes basic requirements for licensing of commercial drivers. In Canada, Transport Canada (TC) has the same authority. Both the DOT and TC establish special requirements for licensing drivers of trucks and other large vehicles. While these are national guidelines, each state or province has the authority to alter them as it deems necessary for its jurisdiction. Some states and provinces require a fire apparatus driver/operator to obtain a commercial driver's license (CDL). Other states have exempted fire service personnel from these licensing requirements. The health and safety officer must be aware of the requirements within the jurisdiction and make sure that the driver/operators are licensed accordingly. If a member should have his license suspended or revoked or have restrictions placed on it, the member must not be allowed to operate fire department vehicles/apparatus.

A commercial driver's license is required for the operation of fuel trucks, passenger buses, and tractor-trailer-type trucks. Uniformed personnel or civilian operators must take the training and pass the CDL test before they can operate these vehicles. Personnel without a CDL should never be allowed to operate these types of vehicles.

Training and Certification

In order to develop an effective vehicle/apparatus safety program, the department has to provide a training program that encompasses driving and operating fire apparatus/vehicles. One recommended driver training course is the National Safety Council's eight-hour defensive driving course (Figure 8.36). The National Safety Council also offers an educational program specifically designed for fire and EMS driver/operators. The program should include instruction on department SOPs relating to apparatus/vehicle operations (emergency and nonemergency) and response. In addition, the Volunteer Firemen's Insurance Services (VFIS) and many national insurance companies provide excellent driver training programs.

Apparatus equipped with engine, transmission, or driveline retarders require specific operator training. Retarders act with the apparatus braking system to slow the apparatus when pressure is released on the accelerator pedal. Depending on the design of the retarder, it lowers the engine speed, downshifts the transmission, or applies a magnetic brake on the drive shaft that slows the apparatus. Use of the retarder creates a high-pitched whine that can exceed allowable noise limits. Some jurisdictions prohibit or restrict the use of retarders. NFPA 1500 requires that driver/operators be provided with training if they are to operate apparatus equipped with retarders.

In addition, the fire department must develop SOPs that require drivers to discontinue the use of manual brake-limiting valves (frequently labeled as a "wet-road/dry-road" switch) and require that the valve/switch remains in the "dry-road" position (Figure 8.37). This type of auxiliary braking system was commonly installed in apparatus built before the mid-1970s, although some more recent

Figure 8.36 The health and safety officer must ensure that driver training is provided for all driver/operators.

Figure 8.37 A typical wet-road/dry-road switch.

vehicles may have the system. The devices were intended to help the driver/operator maintain control of the apparatus on slippery surfaces. This control was accomplished by reducing the air pressure to the front steering axle by 50 percent when the switch was in the slippery or wet-road position. This reduction in pressure prevented the front wheels from locking, allowing the driver/operator to steer the vehicle even when the rear wheels were locked into a skid. In reality, these systems were not overly effective or safe. After the United States government adopted the Federal Motor Vehicle Safety Standard 121 in 1975, few trucks were built with this braking system installed. *IFSTA recommends that apparatus equipped with this switch have it placed in the dry-road position and disconnected.*

Emergency Vehicle Operator's Course

Members who are required to drive and operate a variety of fire department vehicles/apparatus must be properly trained in emergency and nonemergency conditions. The Emergency Vehicle Operator's Course (EVOC) provides the medium for potential driver/operators and current driver/operators to be certified or recertified. Most state fire service training agencies offer this course which coupled with the National Safety Council's Defensive Driving Course is a total of 40 hours of education, training, and certification relating to the operation of fire department vehicles/apparatus. The fire department's training program should clearly outline the training, certification, and recertification requirements for the operation of vehicles/apparatus. The certification and recertification course schedules should be staggered to prevent removing too many companies or driver/operators from service at one time. This training and certification program can be based on NFPA 1002. Personnel assigned to aircraft fire fighting and rescue apparatus must meet the requirements of NFPA 1003, *Standard for Airport Fire Fighter Professional Qualifications.*

 ## Vehicle/Driver Operating Policies and Practices

The development, implementation, and periodic review of SOPs regarding the operational use of all fire department vehicles/apparatus are vital components of the department's risk management plan. Written procedures must clearly define what is expected of personnel in the operation of the department's vehicles/apparatus. The policies include dispatch and response policies, emergency response practices, apparatus placement at incidents, returning from incidents policies, and nonemergency operation policies.

Dispatch and Response Policies

A fire department's response policy should dictate the initial dispatch of equipment for the type of incident. Many incidents require the immediate attention of firefighters but are not life-threatening situations. These situations do not warrant an emergency response that may generate unnecessary risks to firefighters, citizens, and property. There are numerous situations that do require an emergency response. The incident commander, telecommunicator, company officer, battalion officer, or other chief officer always has the opportunity to upgrade or downgrade a response based upon information received or additional case/incident comments. Unit(s) assigned to an incident have the responsibility to respond in the appropriate mode. In some jurisdictions, the telecommunicator has the authority to increase, decrease, or alter the basic response. The health and safety officer should ensure that all personnel are aware of this policy if it applies to the jurisdiction.

Emergency Response Practices

The safe and efficient arrival of apparatus and personnel at an emergency incident are paramount and the first priority. Department emergency response SOPs should include the following guidelines:

- Knowing maximum apparatus speed during response

- Approaching and entering intersections safely

- Crossing railroad grade crossings safely

- Operating in school zones or passing school buses properly

- Using audible and visual warning devices properly

- Providing right-of-way for other responding apparatus appropriately
- Knowing procedures in the event the apparatus is involved in a collision

In addition, it is the responsibility of the company officer to assist the operator in anticipating and identifying hazards while responding to incidents. The officer in charge, however, should not manipulate the emergency equipment, including warning devices when responding to an incident. The apparatus is under the care and control of the driver/operator, who is ultimately responsible for its operation. Altering the emergency equipment may create liability if a mishap were to occur.

Apparatus Placement at Incidents

Apparatus placement at emergency incidents is dictated by the specific circumstances of the incident and the surrounding environment. However, company officers and apparatus driver/operators must be trained in the general guidelines for safe and effective apparatus placement. These guidelines may be in the department's standard operating procedures manual or in the incident management plan. The health and safety officer must participate in the development of these policies and plans to ensure a safe working environment.

For incident control to be achieved efficiently and safely, driver/operators must position apparatus so that its use is maximized. Proper positioning of the apparatus provides a safety barrier that protects the scene, victims, and emergency personnel. Proper positioning also provides a protected work area (Figure 8.38). When positioning apparatus, driver/operators must allow for adequate parking of additional fire department apparatus. Driver/operators must allow enough distance to prevent a moving vehicle from striking and forcing fire apparatus into the work area. Where possible, driver/operators should position apparatus at a 45-degree angle into the curb. This helps to direct motorists around the scene. At intersections or where the incident may be near the middle of the street, two or more sides of the incident may need to be protected. All exposed sides should be blocked if possible. The health and safety officer may also wish to equip each apparatus with safety cones for use around an incident scene. During pump opera-

tion, the pump panel should be positioned at curbside if possible. When laying hose and positioning at a water source, the driver/operator must take the necessary steps to warn motorists of these operations. The incident commander and the health and safety officer or incident safety officer must perform a continuous risk assessment of the scene to ensure the safest working environment possible for firefighters.

For safety considerations, the driver/operator must not position the apparatus under overhead power lines, too close to a potential structural collapse/fire spread, or in the access or egress path of other apparatus (Figure 8.39). The health and safety officer should establish a policy that requires the driver/operator to chock the apparatus wheels when the vehicle is parked at the incident scene (Figure 8.40).

Driver/operator training must also address weight limitations of roadbeds, bridges, or parking structures and incident traffic. Equipment carried in the cab area must be secured as well. Most importantly, the driver/operator must ensure that all

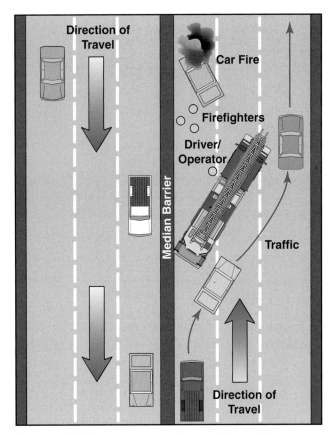

Figure 8.38 The apparatus can be used to provide a barrier between firefighters and oncoming traffic.

Figure 8.39 Driver/operators must use care to avoid tangling the aerial device in overhead wires. *Courtesy of Ron Jeffers.*

Figure 8.40 When the apparatus is parked at an incident, the tires must be chocked.

persons on the apparatus are seated and belted. Visual warning devices must be turned off while returning to quarters.

Should the company receive an alarm while en route from an incident, the officer must take the necessary steps to ensure that the firefighters can safely and properly dress for the type of incident. This situation may require the driver/operator to stop and pull off the road to allow all members to put on the proper protective equipment needed, then be seated and belted.

Figure 8.41 The rearward movement of apparatus should be directed by two firefighters, one equipped with a portable radio.

Operating Apparatus in Reverse

Because a large percentage of collisions occur while apparatus are backing, the health and safety officer should ensure that all personnel are familiar with department policy and trained in the correct procedures for operating the apparatus in reverse. Whenever possible, the driver/operator should simply avoid backing the fire apparatus. If the situation occurs that requires the rearward movement of the vehicle, then one and preferably two firefighters with portable radios should be assigned to direct the driver/operator. It is their duty to clear the way and warn the driver/operator of any obstacles in the path of the apparatus. This very simple procedure can prevent a large percentage of collisions that occur during backing operations (Figure 8.41).

Nonemergency Operation of Apparatus and Vehicles

Policies regarding the nonemergency operation of apparatus and other fire department vehicles must be defined and conveyed to all personnel. These policies include safe driving practices, adhering to all local and state traffic laws, and defensive driving techniques. Driver/operators and company members must be familiar with issues such as the correct rearward movement of the apparatus, correct parking techniques, and unsafe practices such as standing while the apparatus is in motion. These requirements extend to all members of the department who operate department apparatus/vehicles including staff members, inspectors, training personnel, and supply personnel.

In the event that a department vehicle/apparatus is involved in a collision, the driver/operator must be familiar with the protocol for reporting such an event. Department policy usually includes performing the following procedures:

- Tend to injuries.
- Notify dispatch/telecommunications center.
- Call for the immediate supervisor.
- Call for the health and safety officer.
- Survey damage to all vehicles.
- Collect names of witnesses.
- Fill out the appropriate incident reporting form.
- Place the company or unit back into service.

The collision policy also defines the incident investigation process and the required resolution of the incident. Depending on the severity of the incident, the resolution may include remedial training, loss of points, loss of driving privileges, or some other action. The health and safety officer is involved throughout the process and may be called as an expert witness in a legal action resulting from the incident. Therefore, the health and safety officer is responsible for objectively collecting the incident data and analyzing it.

Private Vehicles

When members are allowed to respond to incidents or respond to the fire station in their own vehicles, these operations are regulated by state motor vehicle codes and regulations. The fire department must develop and implement specific SOPs pertaining to the emergency response of private vehicles. These SOPs should be as stringent as those regulating the operation of fire department apparatus.

Allowing members to respond in their private vehicles in emergency mode may or may not affect the licensing requirements. Responding to emergencies in their personal vehicles could impact the members' and the fire department's insurance rates (Figure 8.42).

When private vehicles respond to the incident scene, the fire department must enact provisions to ensure that access to the incident is not blocked or hampered by these vehicles. The fire department may require that members respond to the fire station to limit the number of vehicles that respond to an incident. Areas with limited access can become

Figure 8.42 In some states, volunteers are permitted to equip their personal vehicles with warning lights for responding to emergencies.

quickly congested with vehicles if procedures do not dictate positioning of personal vehicles and other nonemergency vehicles. Staging areas must be established and controlled by personnel assigned the task in the incident management plan.

Jurisdictional Review of Policies

Necessary steps must be taken to ensure that fire department or jurisdictional legal representatives review all SOPs and policies regarding the operation of fire department vehicles/apparatus. The liability of operating fire department vehicles/apparatus on the roadway in a nonemergency mode is reason enough to prompt assistance. Legal representatives should review and approve SOPs regarding response to incidents, licensing requirements, insurance requirements, use of personal vehicles, and operator certification.

◆ Summary

The health and safety officer plays an important role in the management of the fire department's apparatus and vehicle assets. He assists in the development of apparatus specifications, reviews the applicable apparatus safety standards, ensures the periodic safety testing of apparatus, and develops the safety component of the driver/operator training program. He also participates in the investigation of incidents that result in vehicle/apparatus damage or injuries. His knowledge of ergonomics, apparatus standards, and safe apparatus/vehicle operation can provide the fire department with the safest possible apparatus/vehicles within the financial limitations of the governmental body.

9

Incident Scene Safety

This chapter provides information that will assist the reader in meeting the following job performance requirements from NFPA 1521, *Standard for Fire Department Safety Officer,* 1997 edition.

4-1.1 The incident safety officer shall be integrated with the incident management system as a command staff member, as specified in NFPA 1561, *Standard on Fire Department Incident Management System.*

4-1.2 Standard operating procedures shall define criteria for the response or appointment of an incident safety officer. If the incident safety officer is designated by the incident commander, the fire department shall establish criteria for appointment based upon 4-1.1 of this standard.

4-1.3 The incident safety officer and assistant incident safety officer(s) shall be readily identifiable on the incident scene.

4-2.1 The incident safety officer shall monitor conditions, activities, and operations to determine whether they fall within the criteria as defined in the fire department's risk management plan. When the perceived risk(s) is not within these criteria , the incident safety officer shall take action as outlined in Section 2-5.

4-2.2 The incident safety officer shall ensure that the incident commander establishes an incident scene rehabilitation tactical level management unit during emergency operations.

4-2.3 The incident safety officer shall monitor the scene and report the status of conditions, hazards, and risks to the incident commander.

4-2.4 The incident safety officer shall ensure that the fire department's personnel accountability system is being utilized.

4-2.5 The incident commander shall provide the incident safety officer with the incident action plan. The incident safety officer shall provide the incident commander with a risk assessment of incident scene operations.

4-2.6 The incident safety officer shall ensure that established safety zones, collapse zones, hot zone, and other designated hazard areas are communicated to all members present on scene.

4-2.7 The incident safety officer shall evaluate motor vehicle scene traffic hazards and apparatus placement and take appropriate actions to mitigate hazards.

4-2.8 The incident safety officer shall monitor radio transmissions and stay alert to transmission barriers that could result in missed, unclear, or incomplete communication.

4-2.9 The incident safety officer shall communicate to the incident commander the need for assistant incident safety officers due to the need, size, complexity, or duration of the incident.

4-2.10 The incident safety officer shall survey and evaluate the hazards associated with the designation of a landing zone and interface with helicopters.

4-3.1 The incident safety officer shall meet the provisions of Section 4-2 of this standard during fire suppression operations.

4-3.2 The incident safety officer shall ensure that a rapid intervention crew meeting the criteria in Chapter 6 of NFPA 1500, *Standard on Fire Department Occupational Safety and Health Program*, is available and ready for deployment.

4-3.3 Where fire has involved a building or buildings, the incident safety officer shall advise the incident commander of hazards, collapse potential, and any fire extension in such building(s).

4-3.4 The incident safety officer shall evaluate visible smoke and fire conditions and advise the incident commander, tactical level management unit officers, and company officers on the potential for flashover, backdraft, blow-up, or other fire event that could pose a threat to operating teams.

4-3.5 The incident safety officer shall monitor the accessibility of entry and egress of structures and the effect it has on the safety of members conducting interior operations.

4-4.1 The incident safety officer shall meet the provisions of Section 4-2 of this standard during EMS Operations.

4-4.2 The incident safety officer shall ensure compliance with the department's infection control plan and NFPA 1581, *Standard on Fire Department Infection Control Program*, during EMS operations.

4-4.3 The incident safety officer shall ensure that incident scene rehabilitation and critical incident stress management are established as needed at EMS operations, especially mass casualty incidents (MCI).

4-5.1 The incident safety officer shall meet provisions of Section 4-2 of this standard during hazardous materials incidents.

4-5.2 The hazardous materials incident safety officer shall meet the requirements of Chapter 4 of NFPA 472, *Standard for Professional Competence of Responders to Hazardous Materials Incidents.*

4-5.3 The incident safety officer shall attend strategic and tactical planning sessions and provide input on risk assessment and member safety.

4-5.4 The incident safety officer shall ensure that a safety briefing, including an incident action plan and an incident safety plan, is developed and made available to all members on the scene.

4-5.5 The incident safety officer shall ensure that hot, warm, decontamination, and other zone designations are clearly marked and communicated to all members.

4-5.6 The incident safety officer shall meet the incident commander to determine rehabilitation, accountability, or rapid intervention needs. For long-term operations, the incident safety officer shall ensure that food, hygiene facilities, and any other special needs are provided for members.

4-6.1 The incident safety officer shall meet the provisions of Section 4-2 of this standard during special operations incidents. The individual that serves as the incident safety officer for special operations incidents shall have the appropriate education, training, and experience in special operations.

4-6.2 The incident safety officer shall attend strategic and tactical planning sessions and provide input on risk assessment and member safety.

4-6.3 The incident safety officer shall ensure that a safety briefing, including an incident action plan and an incident safety plan, is developed and made available to all members on the scene.

4-6.4 The incident safety officer shall meet the incident commander to determine rehabilitation, accountability, or rapid intervention needs. For long-term operations, the incident safety officer shall ensure that food, hygiene facilities, and any other special needs are provided for members.

4-7.1 Upon notification of a member injury, illness, or exposure, the incident safety officer shall immediately communicate this to the incident commander to ensure that emergency medical care is provided.

4-7.2 The incident safety officer shall initiate the accident investigation procedures as required by the fire department.

4-7.3 In the event of a serious injury, fatality, or other potentially harmful occurrence, the incident safety officer shall request assistance from the health and safety officer.

4-8.1 The incident safety officer shall prepare a written report for the postincident analysis that includes pertinent information about the incident relating to safety and health issues. The incident safety officer shall participate in the postincident analysis.

4-8.2 The incident safety officer shall include information about issues relating to the use of protective clothing and equipment, personnel accountability system, rapid intervention crews, rehabilitation operations, and other issues affecting the safety and welfare of members at the incident scene.

Chapter 9

Incident Scene Safety

Firefighters are at their greatest risks during emergency operations. Whether they are involved in structural fire fighting, water rescue, extrication, or a medical response, firefighters place themselves in harm's way by the mere fact of performing their duties. To ensure that firefighters do not become victims, the health and safety officer defines potential hazards and addresses them in a proactive manner. The proactive approach, as discussed in the previous chapters, includes apparatus safety, equipment maintenance and use policies, protective clothing design and maintenance, health and wellness programs, safety policies and procedures, and facilities design. This chapter addresses the safety-related concerns of the incident scene.

 Incident Safety Officer

During an incident, the incident commander (IC) has the ultimate responsibility for safety-related issues. At a large incident, the safety function may be given to another chief officer, a trained company officer, the health and safety officer, another member of the safety office, a staff officer trained to perform this function, or a combination of these individuals. On smaller departments, the incident commander may find that he has to perform this function along with other duties. Regardless of the type of incident, the role of the incident safety officer must be defined in the fire department's written standard operating procedures (SOPs). NFPA 1521, *Standard for Fire Department Safety Officer*, outlines the role of the incident safety officer and is a good starting point for developing the SOP.

Due to the increasing responsibilities of the health and safety officer, NFPA 1521 has separated the duties of the health and safety officer and the incident safety officer. This separation allows the fire department administration the ability and authority to assign this second function to other qualified personnel. Chapter 1, The Evolution of the Fire Department Safety Officer, outlines the functions and qualifications for the separate positions.

Statistics published in the November/December 1999 issue of the *NFPA Journal* indicate that more firefighters are killed and injured at the incident scene than during any other activity. Using an incident safety officer provides a means of focusing specifically on safety while monitoring conditions, activities, and operations to determine whether appropriate risk management procedures are being followed. Through continual risk assessment, the incident safety officer evaluates and suggests effective tactics that provide a successful outcome of the incident while ensuring the safety of the members operating at the incident (Figure 9.1). Other duties of the incident safety officer as outlined in NFPA 1521 include the following:

- Ensure that incident scene rehabilitation is established.

- Monitor the scene and report the status of conditions, hazards, and risks to the incident commander.

- Ensure that a personnel accountability system is being used.

- Ensure that all personnel understand the incident action plan.

- Provide the incident commander with a risk assessment of the incident action plan.

- Suggest safety zones, collapse zones, a hot zone, and other designated hazard areas.

Figure 9.1 The incident safety officer is responsible for ensuring that potential hazards are recognized and addressed at emergency incidents. *Courtesy of Ron Jeffers.*

Figure 9.2 The incident safety officer is a part of the incident command staff and works closely with them. *Courtesy of Martin Grube.*

- Evaluate motor vehicle traffic hazards.
- Monitor radio transmissions to ensure proper and effective communications.
- Identify the need for additional assistant incident safety officers.
- Evaluate hazards associated with helicopter landings.

For the incident safety officer to be effective, he must have the support of the fire department administration. This supportive atmosphere creates a safety-oriented culture within the department. Written policies and procedures that are based on NFPA 1500, 1521, and 1561 (in addition to other safety standards) provide the documented foundation for this culture. Further, the administration must support the decisions of the incident safety officer. Equally as important, when senior administrative officers arrive on the scene and take command, it is essential that they set the appropriate example and follow their own SOPs (for example, wearing complete protective clothing and an SCBA in the hot zone or areas of potentially IDLH atmospheres).

The success of the incident safety officer requires more than the support of the administration — he must have credibility, good interpersonal skills, and mutual respect for the personnel with whom he works (Figure 9.2). These factors are described as follows:

- *Credibility* — Based on the incident safety officer's knowledge of safety-related issues and hazards found at incident scenes, the incident management system used, and a willingness to enforce safety policy.
- *Interpersonal skills* — Include the ability to remain calm during a crisis, to communicate information clearly and concisely, and to firmly introduce decisions without alienating personnel. (Interpersonal skills can be learned through courses in interpersonal communication normally taught in colleges and universities.)
- *Mutual respect* — Attained when each party recognizes the duties and responsibilities of the other and agrees to work together as a team. The incident safety officer must support the decisions of the incident commander unless the decisions adversely affect personnel safety. The incident commander must be prepared to accept the incident safety officer's recommendations and implement them as part of the incident action plan (Figure 9.3). This may include ceasing an operation assigned by the IC.

◆ Incident Management System

As the complexity of fire fighting operations and other emergency incident scenes has increased

Figure 9.3 The incident commander depends on accurate and timely safety reports from the incident safety officer.

during the last half of the twentieth century, the fire service has attempted to improve its incident scene organization and use of resources. Established by NFPA 1500, the Incident Management System (IMS) is a written standard operating procedure for handling all types of emergency responses. It is also applicable to training exercises and drills that involve hazards similar to those found in actual incidents. NFPA 1500 requires that all members of the department be trained in the Incident Management System and that they apply it during operations. The primary goals of the Incident Management System are to manage all resources (including firefighters and other personnel), ensure effective service delivery, and ensure the safety of all members operating at each incident scene regardless of the size of the incident.

The Incident Management System is designed to be applicable to incidents of all sizes and types. It applies to both small, single-unit incidents that may last a few minutes and complex, large-scale incidents and involve several agencies and many mutual-aid units. The IMS has a number of interactive components that provide the basis for clear communication and effective operations. These components include the following:

- Common terminology
- Modular organization
- Integrated communications
- Unified command structure

- Consolidated action plan
- Manageable span of control
- Predesignated incident facilities
- Comprehensive resource management

The five major operational positions within the IMS are as follows:

- ***Incident commander and command staff*** — The incident commander is responsible for effectively managing the incident. The command staff, which reports directly to the incident commander, includes the incident safety officer, public information officer, and the liaison officer.

- ***Operations*** — The person in charge is responsible for all tactical operations at an incident.

- ***Planning*** — The person in charge is responsible for the collection, evaluation, dispersion, and use of information about the development of an incident and the status of resources.

- ***Logistics*** — The person in charge is responsible for providing facilities, services, and materials for an incident.

- ***Administration/finance*** — The person in charge is responsible for all costs and financial transactions associated with an incident.

This IMS model of organization provides for a unity of command, which is made up of units and personnel that have specific tasks and responsibility (Figure 9.4). Each officer (from the incident

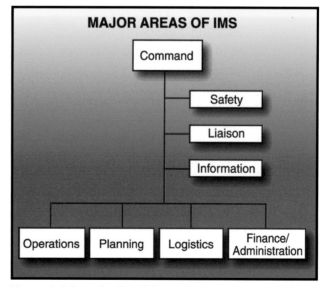

Figure 9.4 Organization of the command staff in the Incident Management System.

commander to the line company officer) has a manageable span of control. This organizational structure makes it much easier for each supervisor to keep track of subordinates and ensure their safety. For more information on this system, see NFPA 1561, the IMS Consortium manuals, or IFSTA's **Fire Department Company Officer** and **Essentials of Fire Fighting** manuals.

In addition to organizing the department's resources, the Incident Management System divides the incident scene into divisions (geographic areas), groups (functional areas of responsibility), and sectors (either geographic or functional areas). These areas are described as follows:

- *Divisions* — Assigned the responsibility of all operations within a defined area. Division assignments begin with Division A located at the front (street address side) of an incident and are assigned clockwise around the scene. If an incident occurs in a single-story structure, the entire interior is designated as the Interior Division. In multistory structures, the division is usually designated by the floor on which it operates (for example, Division 1 for the first floor).

- *Groups* — Based on the type of task required such as forcible entry, ventilation, etc.

- *Sectors* — Either geographic or functional assignments that are equivalent to a division or a group. Each of these units has its own supervisor. The incident safety officer, as part of the command staff, must be aware of the organization of the incident scene. If the operation is large or complex, additional incident safety officers may be assigned to specific divisions, groups, or sectors.

 ## Scene Safety Concerns

To adequately advise the incident commander, the incident safety officer must be knowledgeable in emergency scene management and the potential hazards that different types of incidents pose to firefighters. The first-arriving company officer and the incident safety officer— as well as the incident commander — must be able to assess the incident upon arrival.

As mentioned previously, the Incident Management System is a common organizational structure

Figure 9.5 Personnel who may be assigned the function of incident safety officer should work with the health and safety officer to increase their understanding of safety issues.

designed to aid in the management of resources at emergency scenes. The incident safety officer functions within this organization as part of the command staff and must have expert knowledge of his duties and authority as defined by this system (Figure 9.5). Accurate scene assessment is based on the incident safety officer's knowledge and training in the following areas:

- Fire fighting strategies and tactics
- Fire behavior (discussed in the Fire Behavior Considerations section of this chapter)
- Building construction
- Fire loads
- Forcible entry
- Ventilation
- Evacuation procedures
- SCBA and personal protective equipment usage
- Climatic conditions
- Ladder operations
- Establishing control zones

Fire Fighting Strategies and Tactics

Fire fighting strategies and tactics are the methods employed in the suppression of all types of fires — structural, vehicular, wildland, hazardous materials, or aircraft. Fire fighting strategies and tactics include the placement of apparatus, the use of attack and support hoselines, ventilation procedures, rescue operations, and forcible entry operations. The incident safety officer must understand the various types of tactics used by the department. This understanding enables him to spot the potential hazards created by tactics inconsistent with the IC's incident strategies or by the improper execution of tactics (for example, using opposing hose streams) and to warn the incident commander of such events.

Building Construction

The incident safety officer must have a good working knowledge of building construction and design and the associated fire risks. This knowledge allows him to correctly assess the potential hazards created by such designs as bowstring construction roofs, walls without fire-stops, lightweight wood truss roofs, unsupported exterior walls (may be prone to collapse), or metal structural members (may fail during prolonged exposure to heat or fire) (Figure 9.6).

Fire Loads

Fire loading is the maximum heat that can be produced if all the combustible materials in a given area burn. These materials include the building structure, interior finishes and furniture, and materials stored within the fire area. The incident safety officer must have a working knowledge of target hazards within the jurisdiction such as lumberyards, warehouses, and refineries. Sometimes the decision on whether to conduct offensive or defensive fire fighting operations is based on fire load estimates and fire flow formulas to determine whether available resources can meet fire flow. The incident safety officer must be able to judge the fire load based on the information available and provide the incident commander with an assessment of potential structural failure or fire expansion.

Forcible Entry

The incident safety officer must be familiar with both the forcible entry techniques used by the department and the hazards inherent in this activity. Hazards include broken glass in doors, windows, and skylights and loose and flying debris such as masonry shards, wood splinters, and metal shards. Hazards caused by forcible entry techniques may include muscle strains from improper use of axes or sledgehammers, injuries from the use of axes that "bounce back" after striking a soft surface, or back problems from using a battering ram.

Ventilation

The incident safety officer must be familiar with ventilation principles and the specific techniques used by the department. Improper ventilation can cause fire extension, force heat or smoke onto victims, force interior crews to withdraw, or increase damage to the structure. The incident safety officer must be familiar with the correct use of negative-pressure and positive-pressure ventilation (PPV) fans (Figure 9.7). The incident commander must

Figure 9.6 An understanding of construction techniques can be gained through inspections of new construction sites.

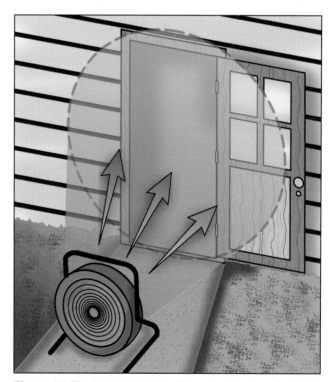

Figure 9.7 The incident safety officer must be familiar with the ventilation principles and techniques used by his department. He should also know the proper and effective way to use positive-pressure ventilation fans.

use the risk management process to determine whether placing firefighters at risk is worth the possible results. The incident safety officer must be prepared to assist in this process by advising the incident commander as needed

Evacuation Procedures

Understanding proper evacuation procedures is important not only in structural fire fighting but also in hazardous materials incidents, wildland fires, and severe weather incidents. The incident safety officer must know the policy for the following actions:

- Ordering an evacuation
- Notifying the people involved
- Effectively using personnel
- Establishing a relocation area with the necessary health and safety provisions

If evacuation involves the use of helicopters, he must be aware of the department's policy on establishing and maintaining landing zones. If firefighters need to be evacuated from an emergency incident, the incident safety officer must perform the following procedures:

- Order the efficient withdrawal of attack crews.
- Advise the incident commander of the need to withdraw.
- Order the use of the rapid accountability system.
- Activate rapid intervention crews (Figure 9.8).

Figure 9.8 The incident safety officer is responsible for ensuring that rapid intervention crews are established. *Courtesy of Ron Jeffers.*

SCBA and Personal Protective Equipment Usage

The incident safety officer must be aware of the policies and standards that mandate the type of protective equipment to be used and the situations requiring its use. He must ensure that all members are properly equipped at the incident and that proper equipment is used. This assurance is especially important with the use of the personal alert safety system. The incident safety officer works with the health and safety officer or the health and safety committee in developing these policies (Figure 9.9).

Climatic Considerations

Weather conditions can have an adverse effect on all types of operations and can affect the health and stamina of fire crews. The incident safety officer must be able to monitor changes in weather conditions and notify the incident commander of potential hazards. Adverse weather conditions include high humidity and temperatures, freezing rain, snow, extreme cold temperatures, and high

Figure 9.9 The incident safety officer makes sure that proper protective equipment is in use by all personnel at the incident. *Courtesy of Mike Nixon.*

wind (Table 9.1). Wind direction can also influence weather conditions and fire behavior. It is important to maintain communication with the dispatch/telecommunication center because it can relay information from the local weather service office (Table 9.2).

Ladder Operations

The incident safety officer must be knowledgeable in the proper operation and placement of both ground ladders and aerial devices. Improper placement of ground ladders and aerial devices can

Table 9.1 Heat Stress Index											
	Air Temperature (Degrees F)										
	70	75	80	85	90	95	100	105	110	115	120
Relative Humidity	Apparent Temperature										
0%	64	69	73	78	83	87	91	95	99	103	107
10%	65	70	75	80	85	90	95	100	105	111	116
20%	66	72	77	82	87	93	99	105	112	120	130
30%	67	73	78	84	90	96	104	113	123	135	148
40%	68	74	79	86	93	101	110	123	137	151	
50%	69	75	81	88	96	107	120	135	150		
60%	70	76	82	90	100	114	132	149			
70%	70	77	85	93	106	124	144				
80%	71	78	86	97	113	136	157				
90%	71	79	88	102	122	150	170				
100%	72	80	91	108	133	166					

Apparent Temperature °F	Danger Category	Injury Threat
BELOW 80°	NONE	LITTLE OR NO DANGER UNDER NORMAL CIRCUMSTANCES
80° – 90°	CAUTION	FATIGUE POSSIBLE IF EXPOSURE IS PROLONGED AND THERE IS PHYSICAL ACTIVITY
91° – 105°	EXTREME CAUTION	HEAT CRAMPS AND HEAT EXHAUSTION POSSIBLE IF EXPOSURE IS PROLONGED AND THERE IS PHYSICAL ACTIVITY
106° – 130°	DANGER	HEAT CRAMPS OR EXHAUSTION LIKELY, HEAT STROKE POSSIBLE IF EXPOSURE IS PROLONGED AND THERE IS PHYSICAL ACTIVITY
ABOVE 130°	EXTREME DANGER	HEAT STROKE IMMINENT!

NOTE: Add 10°F when protective clothing is worn and add 10°F when in direct sunlight.
Information from FEMA Publication FA-114/July 1992.

Table 9.2
Wind Chill Chart

Air Temperature (Degrees F)	Wind Speed (Miles per Hour									
	5	10	15	20	25	30	35	40	45	50
35	33	21	16	12	7	5	3	1	1	0
30	27	16	11	3	0	-2	-4	-4	6	-7
25	21	9	1	-4	-7	-11	-13	-15	-17	-17
20	16	2	-6	-9	-15	-18	-20	-22	-24	-24
15	12	-2	-11	-17	-22	-26	-27	-29	-31	-31
10	7	-9	-18	-24	-29	-33	-35	-36	-38	-38
5	1	-15	-25	-32	-37	-41	-43	-45	-46	-47
0	-6	-22	-33	-40	-45	-49	-52	-54	-54	-56
-5	-11	-27	-40	-46	-52	-56	-60	-62	-63	-63
-10	-15	-31	-45	-52	-58	-63	-67	-69	-70	-70
-15	-20	-38	-51	-60	-67	-70	-72	-76	-78	-79
-20	-26	-45	-60	-68	-75	-78	-83	-87	-87	-88
-25	-31	-52	-65	-76	-83	-87	-90	-94	-94	-96
-30	-35	-58	-70	-81	-89	-94	-98	-101	-101	-103
-35	-41	-64	-78	-88	-96	-101	-105	-107	-108	-110

WIND CHILL TEMPERATURE °F	DANGER
ABOVE -25°	LITTLE DANGER FOR PROPERLY CLOTHED PERSON
-25° / -75°	INCREASING DANGER, FLESH MAY FREEZE
BELOW -75°	GREAT DANGER, FLESH MAY FREEZE IN 30 SECONDS

Information from FEMA Publication FA-114/July 1992.

result in the loss of equipment, damage to apparatus, or injuries to personnel. Ground ladders must be placed to provide a means of egress from as well as entry to upper stories or roofs (Figure 9.10). Ground ladders and aerial apparatus must be placed for maximum effectiveness while avoiding overhead obstructions and power lines.

Establishing Control Zones

Control zones allow for the accounting of victims, efficient use of the personnel accountability system, and prevention of nonemergency personnel from endangering themselves. Establishing con-

trol zones at an incident helps to organize the scene into three manageable areas (Figure 9.11). These areas are as follows:

- *Restricted area or hot zone* — Area where the incident is occurring. Only personnel directly involved in the operation and who are fully equipped with protective clothing and SCBA are allowed into this area.

- *Limited access area or warm zone*—Area immediately outside the hot zone occupied by personnel and equipment that are supporting hot zone personnel. Access to this area is limited to personnel supporting the operation and who are wearing protective clothing and SCBA.

Figure 9.10 Ground ladders must be placed to provide a means of egress as well as entry to upper stories.

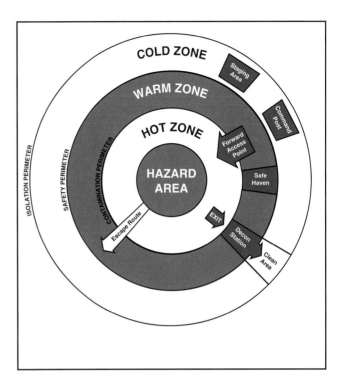

Figure 9.11 The control zones include restricted (hot), limited access (warm), and support (cold).

- *Support area or cold zone* — Outermost ring around the incident where the command post, information officer location, rehab area, and staging area are established. Beyond the cold zone, the outer perimeter is established to provide crowd control and prevent unauthorized personnel from entering.

Other types of protective zones include the *collapse zone*, which is located in the hot zone near walls or structures that may collapse and the *decontamination zone*, which is located in the warm zone where contaminated clothing and equipment can be cleaned or secured. The collapse zone should be equal to one and a half times the height of the building (Figure 9.12). It is established if the structural integrity of the building becomes questionable.

 Fire Behavior Considerations

To be an effective aid to the incident commander, the incident safety officer must be well versed in the

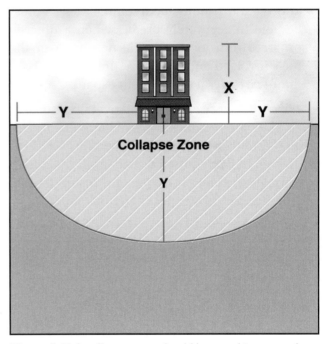

Figure 9.12 A collapse zone should be equal to one and a half times the height of the building.

principles of fire behavior and know how to predict hostile fire events such as flashover and backdraft. He must be trained in the following basic concepts of fire dynamics:

- *Fire tetrahedron* — This model symbolizes the components necessary for fire to exist. It is comprised of oxygen, fuel, heat, and a self-sustained chemical reaction between the first three components (Figure 9.13).

- *Forms and classifications of fuels* — Fuel is the material or substance that is burned in the combustion process. It can be in a solid, liquid, or gaseous form. It may also be classified as Class A (solid ordinary combustibles), Class B (liquid/gas), Class C (energized electrical hazards), or Class D (combustible metals). Knowledge of the classes of fuels and their individual burning characteristics assists the incident safety officer in predicting the potential changes in a fire. He must be able to identify the fuels by their classification symbols.

- *Properties of fuels including flammable ranges* — Understanding the properties of fuels including flammable ranges, ignition point, flash point, Lower Explosive Limit/Upper Explosive Limit (LEL/UEL), and autoignition temperature also helps the incident safety officer provide accurate information to the incident commander. He must be able to accurately predict when a fire will progress from one stage to the next based on the fuels involved and recognize the need to withdraw fire crews if the incident expands.

- *Physical science of fire* — This overall concept of combustion includes types of ignition sources; energy and its application to combustion; heat and temperature theory; heat transfer theory through conduction, convection, and radiation; and the basic principles of combustion.

- *Phases of fire development* — The incident safety officer must be familiar with the phases of fire development — ignition, growth, flashover, fully developed, and decay — to properly read the conditions within a structure fire (Figure 9.14). Within each of these phases, the environment has specific characteristics. Predicting the type of fire fighting tactics to use is based on knowledge of the point the fire has reached and the science of fire behavior. Other situations such as backdraft, smoke explosion, and rollover can occur during the growth and development of a fire.

— Backdraft occurs when air is introduced into an oxygen-starved fire. This situation creates an instant ignition and free-burning state.

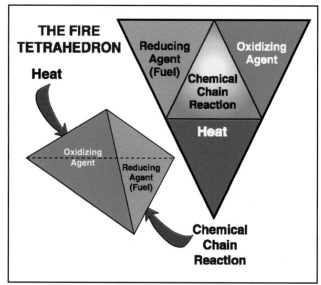

Figure 9.13 Components of the fire tetrahedron.

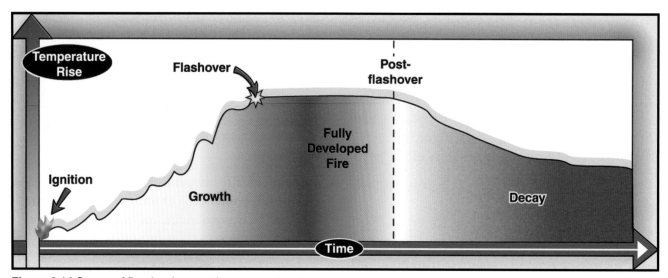

Figure 9.14 Stages of fire development in a compartment.

Figure 9.15 The incident safety officer must ensure that respiratory protection is available and used during the incident.

Figure 9.16 Foam, contaminated water, and hazardous materials must be contained if they pose a threat to the surrounding area.

— Smoke explosion occurs when the unburned products of combustion expand or flash but are not hot enough to ignite.

— Rollover occurs before the flashover phase as small pockets of gas ignite sporadically.

• *Products of combustion* — These are materials released during the burning process. These materials include energy in the form of heat and light, fire gases, particles such as ash and smoke, and liquids. These products not only contribute to the potential for backdraft and flashover but also create a toxic and hazardous atmosphere. This hazardous atmosphere creates the need for continued use of SCBA during the salvage and overhaul processes (Figure 9.15).

 Role and Function of the Incident Safety Officer Prior to an Incident

The incident safety officer's role begins long before arrival at an emergency incident. His role includes involvement in pre-incident planning, development of the department's standard operating procedures, creation and implementation of the Incident Management System, incident safety training, research methodology to gain an understanding of potential hazards, and alarm dispatch and communications procedures.

Pre-Incident Planning

The incident safety officer works with the health and safety officer and field personnel to identify potential high-hazard and high-risk targets in the community. High-hazard occupancies include the following:

• Sites containing large fire loads such as lumberyards and hazardous materials

• Potential sites for large loss of life such as schools, nursing homes, multilevel structures, or structures with multiple below grade levels

• Structures with security bars on the windows

• Doors that impede escape.

The incident safety officer may assist field personnel in developing a strategy for handling various types of incidents involving these target hazards. One consideration that must be addressed is the effect that an incident may have on the environment. Foam, contaminated water, or hazardous materials have to be contained and removed from the site. Pre-incident planning can help to reduce the potential damage to the surrounding area (Figure 9.16). Once these pre-incident plans are created, they must be reviewed and revised periodically to adjust for changes at the site or within the department. Revisions can be based on subsequent inspections or on the postincident analysis report that the incident safety officer prepares following an actual incident.

Standard Operating Procedures

The incident safety officer can assist the health and safety officer and other staff and line personnel in developing the department's SOPs. This familiarity with the SOPs helps the incident safety officer with his performance of emergency incident duties. This involvement also allows his experience from past incidents to promote changes in procedures that created safety hazards.

Incident Management System Implementation

As part of the incident command staff, the incident safety officer must also have an intimate knowledge of the Incident Management System. This knowledge is best attained through involvement in its creation, training or practice, and review and revision processes.

The incident safety officer's role in the IMS includes not only his own training in department IMS operations but also by providing training in safety-related incident issues for other members of the department. Based on postincident analyses and concerns generated by the health and safety committee, safety-related training can be provided to members through the training function.

Research Methodology

In order to gain an understanding of potential hazards, whether created by building construction, fuels, or occupancy types, the incident safety officer must be able to locate and access information regarding them. The incident safety officer needs to be acquainted with the research process that involves determining the topic, working from general to specific, and refining the data. Sources depend on the topic but may include the National Fire Academy Resource Center, Federal Emergency Management Administration, municipal library systems, building code departments, or other agencies. The Internet is also a valuable source for up-to-date material and information. Once the data is compiled from these and other sources, the incident safety officer must analyze it and apply it to potential hazard sites.

Alarm Dispatch and Communications

The incident safety officer is responsible for monitoring all radio transmissions during an incident and ensuring that barriers or obstacles to receiving clear communication are overcome (Figure 9.17).

Figure 9.17 The incident safety officer monitors radio communications at the incident to ensure that all barriers to clear communication are overcome.

The incident safety officer must work to eliminate the following barriers or obstacles to clear communication during an incident:

- Low-charge batteries
- Background noise
- Static
- Bleed-over transmissions
- More than one microphone keyed up at the same time

The incident safety officer must listen for signs of distress in the radio transmissions from fatigued firefighters who are having difficulty talking, firefighters who are reporting trouble situations, or crews who are operating in the wrong area. When preparing a postincident analysis, the incident safety officer must include information that was gained by monitoring radio transmissions and recommend ways to prevent reoccurrence of any problems. The incident safety officer must also report any dispatch procedures that may have resulted in unsafe acts on the part of responding crews.

 ## The Incident Safety Officer's Role at an Incident

When the incident safety officer arrives on the scene of an incident, his first duty is to report to the incident commander (Figure 9.18). The incident commander informs the incident safety officer of the incident action plan. The incident safety officer

Figure 9.18 Upon arrival at the incident, the incident safety officer must report to the command staff.

then provides the IC with a risk assessment of the action plan. The incident safety officer and all personnel performing the function of safety officer must be visibly identified by a vest marked "SAFETY" or by a distinctive helmet. Also, full-protective clothing and equipment are required for all personnel, including safety officers, while working in the hot zone and warm zone.

Conducting Size-Up

Once properly checked in with the IC, the incident safety officer must size up the incident. Size-up is accomplished by doing a complete 360-degree walk-around of the incident scene (sometimes referred to as a "three-sixty") noting the condition of the emergency, the potential hazards to personnel, and the deployment of resources. At structural fires, the incident safety officer notes the type of construction in order to visualize the fuel or fire loading and identify potential weak points. He needs to determine (as nearly as possible) where the fire is located, what is burning, what is the phase or stage of the fire, and what is the potential direction of fire spread. By reading the color, density, and direction of the smoke, the incident safety officer can judge the progress of the fire.

A complete walk-around may not always be possible due to the terrain, fire or hazardous materials spread, traffic congestion, building size, or other congestion situations. However, an attempt should be made to see as much of the scene as quickly as possible. The incident safety officer should also note the need for additional incident safety officers to control remote areas of the scene or to perform

the walk-around while he remains at the command post. The incident safety officer must confirm that a building walk-around has been completed. If a walk-around has not been accomplished, the incident commander and the incident safety officer must initiate a scene assessment walk-around.

Reporting Size-Up to Incident Commander

Returning to the command post, the incident safety officer makes a report on the current conditions to the incident commander. He identifies and prioritizes safety hazards and offers solutions for mitigating them. Identifying safety hazards and reporting them accurately allows the incident safety officer to build credibility with the incident commander. He should also provide the incident commander with any suggestions regarding the safety aspects of the operation. Additional safety personnel may be requested at this time.

Establishing Control Zones

To ensure the effective use of the personnel accountability system, the incident safety officer must establish the separate control zones around the incident. The incident commander may already have done this, but the incident safety officer must ensure that the zones are in place and clearly marked and that personnel are aware of them (Figure 9.19).

Establishing Rapid Intervention Crews

If the operation involves interior structural fire fighting, the incident safety officer ensures that rapid intervention crews are deployed and ready in accordance with NFPA 1500, Chapter 6. He also ensures that entry and egress areas to the structure are accessible.

Additional Duties

The incident safety officer works closely with the incident commander, providing advice on changing conditions, monitoring communications traffic, evaluating vehicular traffic conditions around the incident, monitoring changes in weather conditions, and monitoring the incident action plan and conducting a risk assessment. The incident safety officer continually advises the incident commander

Figure 9.19 To ensure the effective use of the personnel accountability system, the incident safety officer must establish separate control zones using barrier or marker tape. *Courtesy of Ron Jeffers.*

Figure 9.20 If necessary, the incident safety officer establishes a rehabilitation section at the incident. *Courtesy of Ron Jeffers.*

on potential collapse hazards, fire extension, and the potential of flashover or backdraft. If necessary, he establishes a rehabilitation (rehab) section for personnel (Figure 9.20). If the IMS has expanded because of the size of the incident, the rehab section is coordinated with logistics.

 The Incident Safety Officer and the Incident Management System

The incident safety officer is a member of the command staff within the Incident Management System. He is there to advise and support the incident commander, remembering at all times that the incident commander is in charge of the emergency and has final authority. His interaction with the incident commander must be professional, respectful, and ethical. This same interpersonal relationship extends to all section and branch officers at the incident scene, including the officers responsible for divisions, groups, and support functions such as logistics, finance, staging, and rehab.

Interaction with Other Agencies

Because most large-scale operations involve other organizations such as law enforcement; federal, state/provincial, and local emergency management; and public works, the incident safety officer must have a good working relationship with them (Figure 9.21). He must be aware of the authority and responsibility of these various groups. If given the authority for overall incident safety for all responders, he must also communicate that he has this authority to the other agencies. Conflicts in jurisdiction over scene safety can lead to unnecessary losses or injuries and, therefore, must be resolved immediately.

Authority Within the Incident Management System

Both the health and safety officer and the incident safety officer have unique authority within the Incident Management System. In accordance with

Figure 9.21 Large operations may require the incident safety officer to interact with representatives of other agencies or departments. *Courtesy of Chris Mickal.*

NFPA 1521, the health and safety officer has the authority to "cause immediate correction of situations that create an imminent hazard to members." The incident safety officer has the authority to "alter, suspend, or terminate" unsafe activities. In both cases, these actions must be communicated to the incident commander. Examples of the use of this authority include removing a member who is not wearing the required SCBA and personal protective clothing from the hot zone or ordering a crew to withdraw when signs of a flashover are obvious. This authority must be used for good and sufficient reason and must be supported by factual evidence. Whenever possible, altering, suspending, or terminating activities should be done through the IC or division officer. Any activities that are changed by the incident safety officer must be reported to the incident command immediately.

 The Incident Safety Officer and Incident Scene Management

As a member of the incident management command staff, the incident safety officer must be aware of the strategies being used to control an incident. Basically, strategies consist of the steps taken to save lives, reduce property damage, and control the incident. Strategies vary depending on the type of incident involved. For instance, the strategy for rescuing a victim in a confined space is to remove the victim as quickly as possible — stabilizing any injuries — while placing the emergency crews in as little danger as possible.

Strategies in Use at an Incident Scene

When the goal is to extinguish a fire, the incident commander must decide on an offensive, a marginal, or a defensive strategy based on his evaluation of the situation. As the term implies, an *offensive strategy* is an aggressive attack on the fire. This approach is based on the stage of the fire, its location, structural integrity of the building, available resources, etc. The *marginal strategy*, sometimes referred to as the *transitional* or *rescue* strategy focuses on the saving of human life. The marginal strategy is a drastic choice and may result in the violation of such safety procedures as the "two-in/two-out rule." In this strategy, charged hoselines are advanced for the purpose of supporting the rescue and protecting the crews, not for fire extinguishment. A *defensive strategy* is implemented when the stage of the fire is advanced or an interior attack would be too dangerous. In this instance, hose streams are directed at preventing the spread of fire beyond the initial site, and the fire is allowed to burn itself out. These three strategies also apply to hazardous materials incidents and wildland fires. The incident safety officer must be able to advise the incident commander when the situation shifts and a change in strategy is necessary. In any case, offensive, marginal, and defensive strategies shall not be done at the same time.

Tactics Used at an Incident

To accomplish the objective of controlling an incident, the fire department employs tactics. Tactics depend on the type of incident in which the department is involved and the strategy selected to realize the objectives of the incident commander's action plan. Tactics include the placement of hoselines, conducting a type of ventilation, use of search and rescue teams, use of spark patrols to prevent the extension of the fire, or the use of master stream devices. Each of these tactics has inherent safety hazards connected with them.

The incident safety officer must be aware of the tactics in use and ensure compliance with the safety regulations developed for each operation (Figure 9.22). For example, if interior fire fighting operations are planned, a rapid intervention crew must be fully equipped and in place before the attack

Figure 9.22 As the strategy changes at the incident, the incident safety officer must be able to advise the incident commander on new hazards that might arise.

team enters the structure. The incident safety officer must ensure that ventilation techniques do not contribute to the spread or intensity of a fire. He must also monitor the crews in the hot zone to ensure compliance with SCBA and protective clothing policies.

Safety Hazards

When safety hazards are located, the incident safety officer must immediately communicate this fact to the incident commander. If an unsafe hazard or activity is potentially life threatening, he must then exercise his authority to cause the activity to cease or direct firefighters to work around the hazard. He must, at the same time, be aware of the unity of command and that contradictory commands can lead to other problems. When other problems occur, the incident safety officer has to depend on his credibility and the working relationship that he has developed with firefighting personnel to solve them.

Incident Safety Plan

The incident safety officer must develop an incident safety plan. It is based upon the incident commander's strategies and tactics outlined in the incident action plan. The incident safety plan varies according to the type of incident. It is based on the risk management model and takes into consideration the potential hazards and the risks involved.

 ## Challenges for the Incident Safety Officer

The incident safety officer is faced with a multitude of challenges during an emergency incident. Many of these challenges and potential problems may result from outside influences. Regardless of the source, the incident safety officer may be called upon to recommend a solution to such problems.

The following sections describe examples of challenges that may confront the incident safety officer. The best approach is to anticipate the potential problems and attempt to address them in preplanning, standard operating procedures, the incident management system, and the emergency management plan for the jurisdiction.

Communicating Unified Strategies and Tactics

As previously discussed, knowledge of strategies and tactics is important for the incident safety officer. While monitoring the incident scene for changes, the incident safety officer must know how those changes affect the strategies and tactics being employed.

Providing a Risk Assessment

The incident commander ensures that the operations match the strategies developed and implemented by the incident action plan and that safety considerations match the operations. If there are deviations from this plan, a change must occur to prevent loss of control of the incident. A risk assessment ensures that all members operating on the incident scene understand the incident action plan and are conforming to it.

The risk assessment should be conducted periodically throughout the incident, especially when strategic objectives are reached (Figure 9.23). Examples of these objectives or other benchmarks include the following:

• Primary search completed

• Fire knocked down

• Patient extrication completed

• Hazardous material spills or leaks contained

Figure 9.23 Incidents that involve large area structures require ongoing risk assessment of the tactics in use. *Courtesy of Chris Mickal.*

Figure 9.24 The incident safety officer has the responsibility to ensure that safety considerations match the strategy and tactics of the operation. His concerns and analyses must be reported to the incident commander.

- Tactical design
- Routine evaluations and modifications
- Procedural command and management

Responsible decision making and competent tactical objectives ensure the safety and welfare of personnel. As an incident continues, the incident commander must routinely evaluate the incident action plan and modify it as necessary. The incident safety officer must provide the incident commander with accurate assessments of the situation (Figure 9.24). This information allows the incident commander to make the decisions necessary to keep risks at a minimum.

Using Interpersonal Skills and Effective Communication Styles

To effectively communicate with the incident commander and other emergency scene officers, the incident safety officer must have good interpersonal communication skills. He must be able to articulate and express thoughts clearly and concisely. The incident safety officer must also be able to remain calm during stressful situations.

The incident safety officer communicates at an incident scene either by radio or face-to-face interaction (Figure 9.25). The incident safety officer must be prepared to use the radio system, including the department's approved clear-text message system. When possible, face-to-face communication is the most effective type of communication because it reduces the possibility for misunderstandings, pro-

As an objective is met, the incident action plan may change. This change requires another risk assessment of operations.

Managing the Risks

The evaluation and management of risks are the responsibilities of the incident commander. The incident commander must perform a risk analysis to determine what hazards are present, what are the risks to personnel, how can the risks be eliminated or reduced, what are the chances that something may go wrong, and what benefits are gained based upon the strategy employed. By incorporating risk management into the Incident Management System, the basis of emergency incident risk management is built on the following factors:

- Regular assessment of conditions
- Essential decision making

Figure 9.25 Rapid changes in the development of the fire incident require that the incident safety officer is able to communicate effectively with the incident commander to reduce the risks to firefighters. *Courtesy of Chris Mickal.*

Figure 9.26 The incident safety officer works with the incident commander to ensure that the personnel accountability system is implemented.

duces immediate responses, and adds the benefits of nonverbal aspects of communications (facial expressions, gestures, etc.)

Enforcing an Accountability System

The accountability system, which is used to track all personnel at the incident scene, is essential to personnel safety. The incident safety officer is responsible for ensuring that the system is in place and that all members are in compliance. The incident safety officer must have the authority to enforce the use of the system and correct any violations. Working through the incident commander, the incident safety officer may enforce the accountability system and alter tactical operations until all personnel are accounted for (Figure 9.26).

Effectively Using Supporting Sectors/Sections/Branches

The incident safety officer is also responsible for ensuring that other supporting activities are in op-

eration at the scene. These activities include implementing the rapid intervention teams, necessary health and safety logistics functions, and rehabilitation sections. Although he may not take direct control of these functions, he must work closely with the staff and line personnel who are in charge of them. At incidents involving outside agencies, the incident safety officer must be able to assist the incident commander in coordinating activities with them to prevent duplication of efforts. Prior knowledge of these agencies, their functions, and their key personnel is essential to a good working relationship.

Reassessing or Properly Managing Changing Conditions

Failure to properly assess rapidly changing conditions at an incident can be fatal. The incident safety officer must attempt to gather and evaluate the most recent information in order to provide the incident commander with the most accurate information. Changes in wind direction, humidity, temperature, and fire extension and even the loss of daylight need to be taken into consideration. When assessments are incorrect, the incident safety officer must reassess the situation and correct recommendations to fit the new conditions.

Planning for Additional Resources and Mutual Aid

Preplanning for an incident may show that there are insufficient resources to control the incident if certain conditions change. In this case, additional resources including mutual aid may be required. The incident safety officer may suggest the need for additional resources and recommend an increase.

Planning for Protective (Police) and Supportive Resources

Inadequate planning can affect the need for law enforcement (police) protection and logistical support. Incidents that occur during civil unrest may require an increase in police protection for fire fighting forces. To implement this plan, procedures should be in place and communicated to all personnel as part of the standard operating procedures and area emergency management system. The incident safety officer must be able to put these types of plans into operation should the needs arise. Long-term events or those occurring during adverse weather conditions may require additional logistical support. Additional protective clothing to replace contaminated or water-soaked gear must be available. Procedures for acquiring auxiliary power and lighting for night operations must also be in place.

Figure 9.27 The incident safety officer must have a thorough knowledge of the IMS and work closely with all members of the command staff. *Courtesy of Mike Nixon.*

 ## Special Operations

Some of the special operations that the incident safety officer may be called upon to participate in are large-scale operations requiring additional safety officers, hazardous materials operations, emergency medical operations, and specialized rescue operations (Figure 9.27). Each requires that the incident safety officer apply his special knowledge of safety issues to specific types of incidents.

Large-Scale Operations

As emergency incidents expand in size and scope, more resources are required to contain and control them. The incident safety officer may need to request additional safety officers as assistants. Depending on the nature of the emergency, each additional assistant safety officer is assigned a specific task or given responsibility for a given location, function, or duty. At a large-scale fire, assistant safety officers may be assigned to each division providing coverage on all sides of the incident area. They may also be assigned to manage the rehab area or handle the accountability function for off-duty personnel arriving at the scene. The original

incident safety officer then takes on the function of managing the assistants and channeling their reports to the incident commander. They will become his sources of data.

With multiple assistant safety officers on the scene, it is the responsibility of the original incident safety officer to coordinate their safety planning. In essence, each assistant safety officer has his own safety tasks to perform to meet the needs of his individual function. The original incident safety officer must then coordinate these various tasks into the overall plan. This task is accomplished by gathering periodic status reports, monitoring communications, and keeping track of the activities of each section or branch. The incident safety officer may find it advantageous to use a large plot board or marker board to track the assistants and note their reports. This board is similar to the map board frequently employed by the incident commander.

Hazardous Materials Operations

Hazardous materials incidents require the incident safety officer to have special knowledge of the characteristics of various chemicals and substances that constitute this category (Figure 9.28). He must also have a close working relationship with the hazardous materials safety officer, state and federal emergency management personnel, facility/building personnel, and private contractors who are called to assist in cleanup operations. The incident safety officer must establish the control zones, monitor accountability, keep track of changing weather conditions, and ensure that both decontamination and rehabilitation functions are operating. Fully equipped rapid intervention crews must be in place prior to attack crews entering the incident area. The incident safety officer's responsibilities in a hazardous materials incident extend past the completion of the incident. He must ensure that clothing and equipment are properly decontaminated and that the area is either decontaminated or released to an outside contractor who will fulfill the cleaning responsibility.

Emergency Medical Operations

The vast majority of emergency medical incidents in which the fire service responds involve only a single ambulance and fire company. In these cases, the company officer or senior ambulance attendant functions as the incident safety officer. He is guided by his knowledge of the department's standard operating procedures and his knowledge of standards such as NFPA 1581 and OSHA's Title 29 *CFR* 1910.1030.

When emergency medical operations expand in size and involve a large number of casualties, the incident commander should assign or call for the incident safety officer. Mass casualty incidents include (but are not confined to) building collapse, major traffic accidents, aircraft crashes, exposure to hazardous materials leaks, and natural disasters such as tornadoes, earthquakes, and hurricanes. Large numbers of victims must be rescued, triaged, stabilized, and transported (Figure 9.29). The incident safety officer must ensure the safety of emergency personnel at this type of incident and work closely with the triage officer to better communicate needs to the IC. The scope of the incident may require more than one incident safety officer, in which case the procedures described under the Large-Scale Operations section are needed.

Figure 9.28 The incident safety officer must have special knowledge of chemicals and substances when working hazardous materials operations. Establishing rehabilitation and decontamination sections may be part of his duties at such an incident.

Figure 9.29 Incidents involving mass casualties involve rescue, triage, stabilization, and transport of multiple victims. *Courtesy of Ron Jeffers.*

The incident safety officer has to apply his knowledge of rescue techniques, exposure protection, and lifting and carrying techniques to prevent injuries to rescue personnel. His training and the training of response personnel are extremely important to prevent unnecessary injuries to department personnel. The incident safety officer must also be prepared to initiate the critical incident stress management plan for the department and follow up with postincident interviews.

Specialized Rescue Operations

The incident safety officer may be called upon to assist in special operations such as high-angle rescue, confined-space rescue, trench rescue, water rescue, and structural collapse rescue among oth-

Figure 9.30 Water rescue operations, especially in colder climates, pose unique challenges for the incident safety officer. *Courtesy of Angel-Guard Products, Inc.*

ers (Figure 9.30). The incident safety officer needs special training in the fire department's techniques in conducting these types of rescue. He may gain some of this knowledge while working with the health and safety officer and the training officers during the development of the SOPs for these tasks. The incident safety officer must be aware of the types of equipment employed such as hoists, rescue rigging, hard hats, vests, life jackets, and gloves. He must ensure that all safety procedures are followed correctly during the operation.

 Postincident Analyses and Reports

The incident safety officer is responsible for writing a postincident report on all aspects of an incident that involve safety issues and providing it to the health and safety officer and the chief of the department. The incident safety officer also participates in the postincident analysis critique (Figure 9.31). If the incident resulted in the death or injury of a fire department member, he is responsible for investigating the circumstances and reporting the findings. The incident investigation takes the same form as the one described for the health and safety officer in Chapter 2, The Health and Safety Officer as a Risk Manager. Basically, the incident safety officer collects and analyzes the data, reconstructs the incident, and provides recommendations to the health and safety officer or the fire chief.

The postincident report is an important aspect of the incident safety officer's duties. It allows the following safety problems to be corrected:

- Violations of department SOPs
- Poorly defined procedures
- Unforeseen situations
- Training deficiencies

The incident safety officer, working with the health and safety officer, identifies "lessons learned," documents them, and assigns them to the appropriate group or individuals for correction (Figure 9.32). He then tracks the progress to comple-

Figure 9.31 The incident safety officer can use site models to isolate safety problems in the postincident debriefing.

Figure 9.32 The health and safety officer may be present at the incident to work with the incident safety officer in collecting information for the postincident critique.

tion and reevaluates the changes during subsequent incidents. Formal models for applying a "lessons learned" strategy exist in private industry and can be applied to the fire service (Figure 9.33).

Figure 9.33 A postincident critique is an opportunity for the incident safety officer to discuss safety-related issues and recommend corrections.

 ## Summary

The incident safety officer is an essential member of the incident management system. The incident safety officer provides the incident commander with pertinent and accurate reports on safety aspects of the incident, ensures that safety procedures are followed by responding members, coordinates activities with other sections or agencies, and provides the postincident reports on safety issues. Whether this is a function of the health and safety officer, a full-time incident safety officer, or a function of another staff position, the incident safety officer must be well trained in a variety of topics from fire behavior to medical evacuation procedures. He must be capable of communicating distinctly and have good interpersonal skills. The incident safety officer is the key to ensuring as safe a working environment as possible is present at emergency incidents.

A-G

Appendices

Health and Safety Regulations in Canada

In Canada, the provincial and territorial governments regulate health and safety within their own jurisdictions. Although the Canadian federal government has an Occupational Health and Safety (OH&S) agency, it only regulates the activities of some federal government employees. Federal regulations may be found in the Canada Labour Code. Other federal employees are regulated by the OH&S regulations of the province/territory in which they are located. For instance, the Royal Canadian Mounted Police (RCMP) is a federal agency but is required to comply with the provincial/territorial OH&S regulations. Prison guards, who are also federal law enforcement officers, must comply with the federal OH&S regulations and not the local province/territory regulations.

The title of the provincial/territorial agency responsible for administration of OH&S regulations varies. Some of the agencies are Worker's Compensation Board (WCB), Workplace Safety and Insurance Board (WSIB), Worker's Safety and Health, among others. The provincial/territorial agency may be a government agency, a "crown" corporation (government regulated but not funded), or a government board.

Just as the titles of agencies vary from province to province, the OH&S regulations also vary. Regulations only apply in the province/territory enacting them and do not have authority in any other province/territory. Some regulations cover fire service specific subjects such as respiratory protection or protective clothing requirements. Others cover general safety issues for all worker portions of which apply to firefighters.

Provincial/territorial OH&S regulations typically adopt standards or parts of standards within the regulations. Once a standard has been adopted as part of a regulation, the standard must be followed. Standards are only enforceable when they are adopted by the appropriate legislation. The responsibility for enforcement is retained by the provincial/territorial government and not granted by the standard writing agency. This means that Manitoba and not the Canadian Standards Association (CSA) would enforce CSA standards adopted in Manitoba.

Standards may also be listed as reference documents within the OH&S regulations. As a reference document, the cited standard is considered a "recommended practice" that meets the requirements of the OH&S regulation. It is not unusual for provincial/territorial OH&S regulations to adopt or reference National Fire Protection Association (NFPA) standards.

The Standards Council of Canada (SCC) is the verification agency for standards writing agencies. This organization is similar to the American National Standards Institute (ANSI) in the United States. ANSI is also a recognized verification agency in Canada. Standards developed by the NFPA and the National Institute for Occupational Safety and Health (NIOSH) may be adopted in a province/territory because they carry ANSI approval. Standards written by the Canadian Standards Association (CSA) and approved by the SCC may also be adopted or referenced by regulations. Fire service personnel in Canada are urged to consult their provincial OH&S agency to determine the appropriate requirements.

Directory of States with Approved Occupational Safety and Health Plans

Alaska Department of Labor
P.O. Box 21149
1111 W. 8th Street, Room 306
Juneau, Alaska 99802-1149
Ed Flanagan, Commissioner (907) 465-2700
Fax: (907) 465-2784
Alan W. Dwyer, Program Director (907) 465-4855
Fax: (907) 465-3584

Industrial Commission of Arizona
800 W. Washington
Phoenix, Arizona 85007-2922
Larry Etchechury, Director, ICA (602) 542-4411
Fax: (602) 542-1614
Darin Perkins, Program Director (602) 542-5795
Fax: (602) 542-1614

California Department of Industrial Relations
455 Golden Gate Avenue 10th Floor
San Francisco, California 94102
Steve Smith, Director (415) 703-5050
Fax: (415) 703-5114
Dr. John Howard, Chief (415) 703-5100
Fax: (415) 703-5114
Vernita Davidson, Manager, Cal/OSHA Program
Office (415) 703-5177
Fax: (415) 703-5114

Connecticut Department of Labor
200 Folly Brook Boulevard
Wethersfield, Connecticut 06109
James P. Butler, Commissioner (860) 566-5123
Fax: (860) 566-1520

Conn-OSHA
38 Wolcott Hill Road
Wethersfield, Connecticut 06109
Donald Heckler, Director (860) 566-4550
Fax: (860) 566-6916

Hawaii Department of Labor and Industrial
Relations
830 Punchbowl Street
Honolulu, Hawaii 96813
Gil Coloma-Agaran, Director (808) 586-8844
Fax: (808) 586-9099
Jennifer Shishido, Administrator (808) 586-9116
Fax: (808) 586-9104

Indiana Department of Labor
State Office Building
402 West Washington Street, Room W195
Indianapolis, Indiana 46204-2751
John Griffin, Commissioner (317) 232-2378
Fax: (317) 233-3790
John Jones, Deputy Commissioner (317) 232-3325
Fax: (317) 233-3790

Iowa Division of Labor
1000 E. Grand Avenue
Des Moines, Iowa 50319-0209
Byron K. Orton, Commissioner (515) 281-3447
Fax: (515) 242-5144
Mary L. Bryant, Administrator (515) 281-3469
Fax: (515) 281-7995

Kentucky Labor Cabinet
1047 U.S. Highway 127 South, Suite 4
Frankfort, Kentucky 40601
Joe Norsworthy, Secretary (502) 564-3070
Fax: (502) 564-5387
Steven A. Forbes, Federal\State Coordinator
(502) 564-3070 ext. 240 Fax: (502) 564-1682

Maryland Division of Labor and Industry
Department of Labor, Licensing and Regulation
1100 North Eutaw Street, Room 613
Baltimore, Maryland 21201-2206
Kenneth P. Reichard, Commissioner
410) 767-2999 Fax: (410) 767-2300
Ileana O'Brien, Deputy Commissioner
(410) 767-2992 Fax: 767-2003
Keith Goddard, Assistant Commissioner, MOSH
(410) 767-2215 Fax: 767-2003

Michigan Department of Consumer and Industry
    Services
Kathleen M. Wilbur, Director
Bureau of Safety and Regulations
P.O. Box 30643
Lansing, MI 48909-8143
Douglas R. Earle, Director
(517) 322-1814 Fax: (517)322-1775

Minnesota Department of Labor and Industry
443 Lafayette Road
St. Paul, Minnesota 55155
Gretchen B. Maglich, Commissioner
(651) 296-2342 Fax: (651) 282-5405
Michael Houliston, Deputy Commissioner
(651) 296-6529 Fax: (651) 282-5405
Darrell Anderson, Administrative Director, OSHA
    Management Team
(651) 282-5772 Fax: (651) 297-2527

Nevada Division of Industrial Relations
400 West King Street, Suite 400
Carson City, Nevada 89703
Roger Bremmer, Administrator (775) 687-3032
Fax: (775) 687-6305

Occupational Safety and Health Enforcement
    Section (OSHES)
1301 N. Green Valley Parkway
Henderson, Nevada 89014

John Wiles, Acting Chief Administrative Officer
(702) 486-9044
Fax: (702) 990-0358
[Las Vegas (702) 687-5240]

New Mexico Environment Department
1190 St. Francis Drive
P.O. Box 26110
Santa Fe, New Mexico 87502
Peter Maggiore, Secretary (505) 827-2850
Fax: (505) 827-2836
Sam A. Rogers, Chief (505) 827-4230
Fax: (505) 827-4422

New York Department of Labor
W. Averell Harriman State Office Building 12,
    Room 500
Albany, NY 12240
James McGowan, Commissioner (518) 457-2741
Fax: (518) 457-6908
Richard Cucolo, Division Director (518) 457-3518
Fax: (518) 457-6908

North Carolina Department of Labor
4 West Edenton Street
Raleigh, North Carolina 27601-1092
Harry Payne, Commissioner (919) 807-2900
Fax: (919) 807-2855
Robert K. Andrews, Jr., Deputy Commissioner,
    OSH Director
(919) 807-2861 Fax: (919) 807-2855
Kevin Beauregard, OSH Assistant Director
(919) 807-2863 Fax: (919) 807-2856

Oregon Occupational Safety and Health Division
Department of Consumer & Business Services
350 Winter Street, NE, Room 430
Salem, Oregon 97310-0220
Peter DeLuca, Administrator (503) 378-3272
Fax: (503) 947-7461
David Sparks, Deputy Administrator for Policy
    (503) 378-3272 Fax: (503)947-7461
Michele Patterson, Deputy Administrator for
    Operations
(503) 378-3272 Fax: (503) 947-7461

Puerto Rico Department of Labor and Human
  Resources
Prudencio Rivera Martinez Building
505 Munoz Rivera Avenue
Hato Rey, Puerto Rico 00918
Aurea L. Gonzalez-Rios, Secretary (787) 754-2119
Fax: (787) 753-9550
Ana Lopez, Assistant Secretary
(787) 754-2119/2171 Fax: (787) 767-6051

South Carolina Department of Labor, Licensing,
  and Regulation
Koger Office Park, Kingstree Building
10 Centerview Drive
PO Box 11329
Columbia, South Carolina 29211
Rita McKinney, Director (803) 896-4300
Fax: (803) 896-4393
William Lybrand, Program Director
(803) 734-9644 Fax: (803) 734-9772

Tennessee Department of Labor
710 James Robertson Parkway
Nashville, Tennessee 37243-0659
Michael E. Magill, Commissioner (615) 741-2582
Fax: (615) 741-5078
John Winkler, Acting Program Director
(615) 741-2793 Fax: (615) 741-3325

Utah Labor Commission
160 East 300 South, 3rd Floor
PO Box 146650
Salt Lake City, Utah 84114-6650
R. Lee Ellertson, Commissioner (801) 530-6901
Fax: (801) 530-7906
Jay W. Bagley, Administrator (801) 530-6898
Fax: (801) 530-6390

Vermont Department of Labor and Industry
National Life Building - Drawer 20
Montpelier, Vermont 05620-3401
Tasha Wallis, Commissioner (802) 828-2288
Fax: (802) 828-2748
Robert McLeod, Project Manager (802) 828-2765
Fax: (802) 828-2195

Virgin Islands Department of Labor
2203 Church Street
Christiansted, St. Croix, Virgin Islands
00820-4660
Sonia Jacobs-Dow, Commissioner (340) 773-1990
Fax: (340) 773-1858
Marcelle Heywood, Program Director
(340) 772-1315 Fax: (340) 772-4323

Virginia Department of Labor and Industry
Powers-Taylor Building
13 South 13th Street
Richmond, Virginia 23219
Jeffrey Brown, Commissioner (804) 786-2377
Fax: (804) 371-6524
Jay Withrow, Director, Office of Legal Support
(804) 786-9873 Fax: (804) 786-8418

Washington Department of Labor and Industries
General Administration Building
PO Box 44001
Olympia, Washington 98504-4001
Gary Moore, Director (360) 902-4200
Fax: (360) 902-4202
Michael Silverstein, Assistant Director
[PO Box 44600] (360) 902-5495
Fax: (360) 902-5529
Steve Cant, Program Manager, Federal-State
  Operations [PO Box 44600]
(360) 902-5430 Fax: (360) 902-5529

Wyoming Department of Employment
Workers' Safety and Compensation Division
Herschler Building, 2nd Floor East
122 West 25th Street
Cheyenne, Wyoming 82002
Stephan R. Foster, Safety Administrator
(307) 777-7786 Fax: (307) 777-3646

Applicable Codes and Standards

NOTE: This is not a complete list of applicable standards, regulations, and codes. For instance, the Department of Transportation, the United States Coast Guard, the Compressed Gas Association, and the American National Standards Institute issue documents that affect the fire service. In Canada, each province has developed its own occupational safety and health regulations. This list is intended to be a starting point for the health and safety officer.

Standard or Code	Description
OSHA 29 CFR 1910.146	*Permit-required confined spaces*
OSHA 29 CFR 1910.134	*Respiratory protection*
OSHA 29 CFR 1910.120	*Hazardous waste operations & emergency response*
OSHA 29 CFR 1910.156, Subpart L, Appendix A	*Fire protection*
OSHA 29 CFR 1910.1030	*Bloodborne pathogens*
OSHA 29 CFR 1910.1200	*Hazard communication*
OSHA 29 CFR 1910.1926, Subpart P	*Excavations, trenching operations*
OSHA 29 CFR 1910.136	*Occupational foot protection*
OSHA 29 CFR 1910.38	*Employee emergency plans and fire prevention plans*
OSHA 29 CFR 1910.157	*Portable fire extinguishers*
OSHA 29 CFR 1910.147	*The control of hazardous energy (lockout/tagout)*
OSHA 29 CFR 1910.331-335	*Electrical safety-related work practices*
NFPA 1500	*Fire department occupational safety and health program*
NFPA 1521	*Fire department safety officer*
NFPA 1561	*Fire department incident management system*
NFPA 1581	*Fire department infection control program*
NFPA 1582	*Medical requirements for fire fighters*
NFPA 10	*Portable fire extinguishers*
NFPA 101®	*Life Safety® Code*
NFPA 472	*Professional competence of responders to hazardous materials incidents*
NFPA 473	*Competencies for EMS personnel responding to hazardous materials incidents*
NFPA 600	*Industrial fire brigades*

NFPA 1001	*Fire fighter professional qualifications*
NFPA 1002	*Fire apparatus driver/operator professional qualifications*
NFPA 1003	*Airport fire fighter professional qualifications*
NFPA 1021	*Fire officer professional qualifications*
NFPA 1041	*Fire service instructor professional qualifications*
NFPA 1403	*Live fire training evolutions*
NFPA 1901	*Automotive fire apparatus*
NFPA 1911	*Service tests of fire pump systems on fire apparatus*
NFPA 1914	*Testing fire department aerial devices*
NFPA 1931	*Design of & design verification tests for fire department ground ladders*
NFPA 1932	*Use, maintenance, and service testing of fire department ground ladders*
NFPA 1961	*Fire hose*
NFPA 1962	*Care, use, and service testing of fire hose including couplings and nozzles*
NFPA 1964	*Spray nozzles*
NFPA 1971	*Protective ensemble for structural fire fighting*
NFPA 1972	*Helmets for structural fire fighting*
NFPA 1973	*Gloves for structural fire fighting*
NFPA 1974	*Protective footwear for structural fire fighting*
NFPA 1975	*Station/work uniforms for fire and emergency services*
NFPA 1976	*Protective ensemble for proximity fire fighting*
NFPA 1981	*Open-circuit self-contained breathing apparatus (SCBA) for fire fighters*
NFPA 1982	*Personal alert safety systems (PASS)*
NFPA 1983	*Fire service life safety rope and system components*
NFPA 1991	*Vapor-protective ensembles for hazardous chemical emergencies*
NFPA 1992	*Liquid splash-protective clothing for hazardous chemicals*
NFPA 1993	*Support function protective clothing for hazardous chemical operations*
NFPA 1999	*Protective clothing for emergency medical operations*
NFPA 1404	*Fire department self-contained breathing apparatus program*
NFPA 1410	*Training for initial emergency scene operations*
NFPA 1470	*Search and rescue training for structural collapse incidents*
WCB 296/97 part 31	*Workers Compensation Board, British Columbia, Canada, OSHA,*
Locally Adopted Codes	*Uniform building and fire codes*

Risk Management Formulas for Calculating Frequency and Severity

The following formulas may be used to calculate the frequency or incident rate and the severity of incidents.

OSHA calculates the frequency (incident rate) as follows:
$$N/EH \times 200,000 = IR$$

N = number of injuries and/or illnesses
EH = total hours worked by all employees during the calendar year
200,000 = base for 100 full-time equivalent employees (provides standardization between agencies and companies)
IR = incident rate

OSHA calculates the severity as follows:
$$LWD/EH \times 200,000 = S$$

LWD = loss work days
EH = total hours worked by all employees during the calendar year
200,000 = base for 100 full-time equivalent employees
S = severity rate

Another method is to assign values to the frequency and severity in the following formula:
$$R = S \times IR$$
R = risk
S = severity
IR = incident rate

Assessment of Severity

8. Extreme	Multiple deaths or widespread destruction may result from hazard.
7. Very High	Potential death or injury or severe financial loss may result.
6. High	Permanent disabling injury may result.
5. Serious	Loss time injury greater than 28 days or considerable financial loss.
4. Moderate	Loss time injury of 4 to 28 days or moderate financial loss.
3. Minor	Loss time injury up to 3 days.
2. Slight	Minor injury resulting in no loss of time or slight financial loss.
1. Minimal	No loss of time injury or financial loss to organization.

Assessment of Incident Rate

7. Frequent	Occurs weekly.
6. Very Likely	Occurs once every few months.
5. Likely	Occurs about once a year.
4. Occasional	Occurs annually in the United States.
3. Rare	Occurs every 10 to 30 years.
2. Exceptional	Occurs every 10 to 30 years in the United States.
1. Unlikely	May occur once in 10,000 years within the global fire service.

Once the frequency and severity are known, the risk can be plotted on a chart such as the one that follows:

High Frequency

	Area of highest priority
Low Severity	High Severity
Area of lowest priority	

Low Frequency

The U. S. Department of Labor requires that job-related injury and illness reports and logs be maintained by the employer and filed with the federal government periodically. Form 300, *OSHA Injury and Illness Log and Summary*, and Form 301, *OSHA Injury and Illness Incident Record*, are reproduced below and may also be found on the OSHA web site at www.OSHA.gov. The government does allow the employer to use forms other than these as long as the forms are readable, legible, and contain all the information found on the government form. The forms may also be computerized and the information kept in a computer-based system.

OSHA Injury and Illness Log and Summary

Public Law 91-596 and 29 CFR 1904 require you to:

☐ Enter all recordable occupational injuries and illnesses. (See instructions on back.)
☐ Update and retain completed form for three years.
Failure to complete, update and post can result in the issuance of citations and penalties.

U.S. Department of Labor
Occupational Safety and Health Administration

Form approved O.M.B. No. 1218-0000
See O.M.B. disclosure statement on back.

This form is not an insurance form. Cases listed below are not necessarily eligible for Workers' Compensation or other insurance. Listing a case below does not necessarily mean that the employer or worker was at fault or that an OSHA Standard was violated.

Establishment Name

Establishment Address

Mailing Address if different

For calendar year _____

Page ____ of ____

Industry description and Standard Industrial Classification (SIC) if known (*e.g. Manufacture of motor truck trailers, SIC 3715*)

A. Employee's Name (e.g. Doe, Jane B.)	CASE IDENTIFICATION				CASE DESCRIPTION	CASE CLASSIFICATION *(Check only one)*					OTHER
	B. Case Number (*e.g. 1, 2,3*)	C. Date of injury or illness (m/d)	D. Department and location where event occurred (*e.g. loading dock north end*)	E. Regular job title (*e.g. Welder*)	F. Description of injury or illness; part(s) of body affected, and object/substance which directly injured or made employee ill (*e.g. Second degree burns on right forearm from acetylene torch*)	G. Death	H. Involving Days Away	I. Without Days Away			J. Employer Use
						(X)	(X)	(# Days) — Restricted Work Activity (X)	Other (X)		
						☐	☐	☐	☐		
						☐	☐	☐	☐		
						☐	☐	☐	☐		
						☐	☐	☐	☐		
						☐	☐	☐	☐		
						☐	☐	☐	☐		
						☐	☐	☐	☐		
						☐	☐	☐	☐		

YEAR END SUMMARY
Complete the year end portion of this form, even if there were no cases during the year. **Fold along line to the right and post this form from February 1 to January 31 where employees can read it.**

Employees, former employees, and their representatives have the right to review all OSHA Injury and Illness Records, in their entirety, for this establishment.

Year end totals ____ ___ ____ ___ ___

Annual average number of employees _____

Total hours worked by all employees _____

I have examined this Log and Summary and certify its accuracy and completeness **X**_____
(Responsible Company Official)

Title _____ Phone (___)_____ Date __/__/__

Knowingly falsifying this document can result in fine, imprisonment, or both. Draft OSHA Form 300 (10/95)

OSHA Injury and Illness Incident Record

Public Law 91-596 and 29 CFR 1904 require you to update and retain completed form for three years.

Failure to complete this form can result in the issuance of citations and penalties.

Employees, former employees, and their representatives have the right to review all OSHA Injury and Illness Records, in their entirety, for this establishment.

This form is not an insurance form. Cases listed below are not necessarily eligible for Workers' Compensation or other insurance. Listing a case below does not necessarily mean that the employer or worker was at fault or that an OSHA Standard was violated.

U.S. Department of Labor
Occupational Safety and Health Administration

Form approved O.M.B. No. 1218-0000
See O.M.B. disclosure statement on back.

_____ Case number from OSHA Form 300

Employee

1. Last name   First name   MI

2. Male ☐   Female ☐   3. Date of birth   / /

4. Home address

5. Date hired   / /

Health Care Provider

6. Name of health care provider

7. If treatment off-site, facility name and address

8. Hospitalized overnight as in-patient?
(If emergency room only, mark "no")   yes ☐   no ☐

Employer Use (Optional)

Completed by

Name   Title

Phone (   )   Date

Illness or Injury

9. Specific injury or illness
    (e.g. Second degree burn or Toxic hepatitis)

10. Body part(s) affected (e.g. Lower right forearm)

11. Date of injury or illness: / /   12. If employee died, date of death   / /

13. If the case involved days away from work or restricted work
    activity, enter the date the employee returned to work at full capacity: / /

14. Time of event: 15. Time employee began work:
    (Specify a.m. or p.m.)   (Specify a.m. or p.m.)

16. All equipment, materials, or chemicals employee was using when the event occurred.
    (e.g. Acetylene cutting torch, metal plate)

17. Specify activity the employee was engaged in when the event occurred
    (e.g. Cutting metal plate for flooring) Indicate if activity was part of normal job duties.

18. How injury or illness occurred. Describe the sequence of events and include
    any objects or substances that directly injured or made the employee ill.
    (e.g. Worker stepped back to inspect work and slipped on some scrap metal.
    As she fell, worker brushed against the hot metal)

Draft OSHA Form 301 (10/95)

Model Incident Safety Plan

Incident safety plans will vary depending on the type of emergency incident. The department should prepare plans or checklists for each type of incident and keep copies in the IMS resource manual.

Emergency Services Department
Incident Safety Plan

Date: _____   Incident Tracking # _____

Time of Dispatch: _____   Incident Safety Officer: _____

A. Site Assessment

Address: _____   Incident Type: _____

Weather Conditions:

    Wind Direction: _____   Wind Speed: _____

    Temperature: _____   Heat Index: _____

                         Wind Chill: _____

    Precipitation: None [ ]
                         Mist   [ ]
                         Rain   [ ]
                         Sleet [ ]
                         Snow [ ]

Terrain Conditions: _____

Site Accessibility: _____

B. Incident Management System

Location of Command Post: _____

Initial Response: _____

Secondary Response: _____

Deployment by Division: A _____   1 _____
                                         B _____   2 _____
                                         C _____   3 _____
                                         D _____   4 _____

Location of Staging Area: _____

Location of Rehabilitation Area: _____

Assignment of Additional ISOs:
A _____   1 _____
B _____   2 _____
C _____   3 _____
D _____   4 _____
Rehab _____

C. Hazardous Conditions
Type of Condition (structure fire, confined space rescue): _____

Risk Assessment: Severe [ ]   Moderate [ ]   Low [ ]

Probability of Increase of Risk: High [ ]   Medium [ ]   Low [ ]

D. Initial Response
Strategy: Offensive [ ]   Marginal [ ]   Defensive [ ]

Risk Assessment of Strategy: _____

Recommended Changes to Strategy: _____

Tactical Deployment: _____

Risk Assessment of Tactics: _____

Recommended Changes to Tactics: _____

Duration of Incident at Time of Assessment: _____

Primary Goal: _____

Secondary Goal: _____

E. Safety Considerations
Personnel Accountability System Implemented: _____
    Officer in Charge of PAS: _____

Control Zones Established and Marked: _____
    Officer Responsible for Control Zones: _____

Rapid Intervention Crews Established: _____
    Officer in Charge of RIC: _____

Rehabilitation/EMS Support Established: _____
    Officer in Charge of Rehab/EMS: _____

Additional Recommended Changes:

F. Incident Site Plan

Use this space to sketch site plan indicating IMS deployment, wind direction, access, terrain, geographic site, etc.

Sample Fire Department Facility Inspection Form

Facility _____ Date _____

Address _____ Phone _____

Officer in Charge: _____

Inspecting Officer: _____

Check appropriate condition, and list corrective action under comments field:

AREA	EXCELLENT	GOOD	FAIL	COMMENTS
Apparatus room				
PPE storage				
Equipment storage				
Disinfecting area				
Cleaning area				
Watch booth				
Dayroom				
Kitchen				
Classroom				
Officer's room				
Sleeping area				
Bathroom/toilet				
Locker room				
Utility/laundry				
Mechanical room				
Electrical room				
Miscellaneous areas				
Building exterior				

Date range hood extinguishing system last inspected _____
Date fire extinguishers last inspected _____
Date fire alarm system last inspected _____

Recommendations

_____

_____

_____

_____

Corrections to be completed within _____ days.

Signature Line _____ Officer in Charge

Signature Line _____ Health and Safety Officer

A

Accident — *See* Incident.

Accident Investigation — *See* Incident Investigation.

Accident Prevention Program — *See* Incident Prevention Program.

American National Standards Institute (ANSI) — Voluntary standards-setting organization that examines and certifies existing standards and creates new standards.

Americans with Disabilities Act (ADA) — U. S. legislation that establishes guidelines for facility accessibility and hiring practices for persons with disabilities.

Authority Having Jurisdiction (AHJ) — Organization, office, or individual responsible for approving equipment, installations, or procedures.

B

Bloodborne Pathogens — Pathogenic microorganisms that are present in human blood and can cause disease in humans. These pathogens include (but are not limited to) hepatitis B virus (HBV) and human immunodeficiency virus (HIV).

C

Canadian Standards Association (CSA) — Canadian standards writing organization.

Candidate Physical Ability Test (CPAT) — Applicant testing program developed by the International Association of Fire Fighters (IAFF) and the International Association of Fire Chiefs (IAFC). Intended to ensure that applicants can perform the tasks required of a firefighter.

Case Law — Law that results from a legal precedent or a judicial decision. Used as rules in future determinations of similar cases. The impact of case law can be widespread but can also change depending on the decisions of the court in which subsequent cases are heard.

Category A Medical Conditions — Medical condition that keeps a person from performing the duties of a firefighter in training or emergency operations. Such conditions may present a significant risk to the individual or other personnel.

Category B Medical Conditions — Medical condition (depending on its severity) that keeps a person from performing the duties of a firefighter in training or emergency operations. Such conditions may present a significant risk to the individual or other personnel.

Center for Disease Control (CDC) — U. S. government agency for the collection and analysis of data regarding disease and health trends.

Code — Body of law established either by legislative or administrative agencies with rule-making authority. It is designed to regulate as completely as possible the subject to which it relates.

Code of Federal Regulations (CFR) — U. S. government regulations that relate to various topics such as the use of respiratory protection or heavy equipment operation.

Compressed Gas Association (CGA) — Association that writes standards relating to compressed gasses.

Control Zones — System of barriers surrounding designated areas at emergency incident scenes that are intended to limit the number of persons exposed to the hazard and to facilitate its mitigation. At a major incident there are three zones — restricted (hot), limited access (warm), and support (cold).

Critical Incident Stress Debriefing Team — Trained personnel who perform post-incident debriefing of members who have been exposed to any type of psychological/emotional trauma.

Critical Incident Stress Management — Pre- and post-incident education and counseling designed to minimize the effects of psychological/emotional trauma on personnel who are directly involved with victims suffering from particularly gruesome or horrific injuries or death.

D

Defensive Strategy or Mode — Commitment of a fire department's resources to protect exposures when the fire has progressed to a point where an offensive attack is not effective.

Department of Labor — Administrative body of the executive branch of the state/provincial or federal government responsible for labor policy, regulation, and enforcement.

Department of Transportation — Administrative body of the executive branch of the state/provincial or federal government responsible for transportation policy, regulation, and enforcement.

E

Electrically Powered Tools — Power tools that require electricity in order to operate. Tools may be hand held or stationary and powered by battery or AC/DC electrical systems.

Emergency Vehicle Maintenance Certification Program — Certification program for maintenance personnel assigned to work on emergency vehicles and apparatus.

Emergency Vehicle Operators Course — Training program for emergency vehicle operators in the safe handling of vehicles under emergency conditions.

Employee Assistance Program (EAP) — Program designed to provide employees with counseling services ranging from financial counseling to substance abuse programs and domestic violence prevention.

Entry Level Clothing — Type of protective apparel that enables a firefighter to contact flames for a short time. Due to its weight and expense, it is seldom used.

Environmental Protection Agency (EPA) — U. S. government agency that regulates potential hazards to the environment such as water or air pollution.

Equal Opportunity Commission (EEOC) — U. S. government agency that reviews and enforces fair hiring and promotional standards.

Ergonomics — Applied science of equipment and workplace design intended to maximize productivity by reducing operator fatigue and discomfort.

F

Factory Mutual (FM) Approved — Certification of equipment by this independent research and testing laboratory. Equipment is approved only for the specific use for which it is tested.

Fire Department Physician — Officially designated physician who is responsible for guiding, directing, and advising the members of the department on matters of health, fitness, and suitability for duty as defined and required by NFPA 1500.

Fire Fighting Strategy — Overall plan for incident attack and control.

Fire Industry Equipment Research Organization (F.I.E.R.O.) — Voluntary nonprofit organization consisting of industry and fire service personnel. The organization's goal is to provide improvements in the protective clothing and equipment used by the fire service.

Fire Tetrahedron — Model for the four elements required to have a fire. The four sides represent fuel, heat, oxygen, and chemical chain reaction.

G

General Services Administration — U. S. Congressional agency that oversees the operation of government and advises Congress on expenditures of funds.

Ground Fault Interrupter (GFI) — Electrical device designed to discontinue the flow of electricity when grounding occurs. In the event of a short circuit in the device, it prevents electrocution of people in contact with the electrical device.

Guide — Instrument that provides direction or guiding information. Guides do not have the force of law but may provide the basis for what is reasonable in cases of negligence.

H

Health and Safety Committee — Internal fire department committee established to advise the department administration on matters relating to health and safety as defined and required by NFPA 1500.

Health and Safety Officer — Member of the fire department assigned and authorized by the fire chief as the manager of the health and safety program and who performs the duties, functions, and responsibilities specified by NFPA 1521. This individual must meet the qualifications (or approved equivalent) of this standard.

Heat Stress Index — Index or chart that indicates the apparent temperature of the atmosphere when the actual air temperature is effected by the relative humidity in the air. Provides a guide for personnel operating in high- temperature/high-humidity environments.

Heating Ventilation Air Conditioning Systems (HVAC) — Environmental control systems within a structure designed to provide a constant temperature and provide adequate filtered changes in the air in the building.

High Efficiency Particulate Air (HEPA) Filter — Respiratory protection filter designed and certified to protect the user from particulates in the air. The HEPA filter must be at least 99.97% efficient in removing monodisperse particles of 0.3 micrometers in diameter.

Hydraulically Powered Tools — Tools powered by hydraulic pressure. The power source may be a closed circuit system containing hydraulic oil or an open system of water passing through the tool or unit.

Hydrostatic Testing — Testing method used to check the integrity of pressure vessels such as SCBA air cylinders.

I

Incident — For the purpose of this manual, an incident is defined as an unplanned, uncontrolled event resulting from unsafe acts and/or unsafe occupational conditions, either of which result in injury, death, or property damage. Although still used by the NFPA and other organizations, the term *accident* is being replaced by incident or mishap in contemporary usage.

Incident Commander — Person responsible for all operations at an emergency incident.

Incident Investigation — Act of investigating or gathering data to determine the factors that contributed to a fatality, injury, or damage to department property. Also referred to as Accident Investigation.

Incident Management System (IMS) — Standardized system by which facilities, equipment, personnel, procedures, and communications are organized to operate within a common organizational structure designed to aid in the management of resources at emergency incidents. Formerly called the Incident Command System.

Incident Prevention Program — Fire department program that provides instruction in safe work practices for both emergency and nonemergency operations for all members of the department. Also called Accident Prevention Program.

Incident Safety Officer — Individual appointed to respond to an incident scene or assigned at an incident scene by the incident commander to perform duties and responsibilities specified in NFPA 1521. This individual must meet the qualifications of (or equivalent to) this standard.

L

Level A Protection — EPA designation for protective clothing to be worn when the highest level of respiratory, skin, eye, and mucous membrane protection is needed. Consists of positive pressure self-contained breathing apparatus and fully encapsulating chemical resistant suit.

Level B Protection — EPA designation for protective clothing to be worn when the highest level of respiratory protection is needed with a lesser level of skin and eye protection. Consists of positive pressure self-contained breathing apparatus and a chemical resistant overall or splash suit.

Level C Protection — EPA designation for protective clothing to be worn when the type of airborne substance is known, the concentration measured, criteria for using air purifying respirators is met, and skin and eye exposure are unlikely. Consists of air purifying respirator and chemical resistant coverall or splash suit.

Level D Protection — EPA designation for protective clothing to be worn when respiratory or skin hazards do not exist. This is primarily the work uniform.

Lifelines — *See* Life Safety Line or Rope.

Life Safety Harness — Safety harness worn by an individual to prevent falling or to use during rappelling operations. Divided into three classes as defined in the text.

Life Safety Line or Rope System Components — Components used with life safety line and harness including carabiners, snap-links, pulleys, rings, and other devices.

Life Safety Line or Ropes — Rope dedicated solely for the purpose of supporting people during rescue, fire fighting, other emergency operations, or during training evaluations.

Live Fire or Burn Exercises — Training exercises that involve the use of an unconfined open flame or fire in a structure or other combustibles to provide a controlled burning environment.

M

Marginal Strategy — Fire fighting strategy that focuses on the saving of human life. Sometimes referred to as "transitional" or "rescue" strategy. Implementation may result in the violation of safety procedures.

Material Safety Data Sheet (MSDS) — A form provided by the manufacturer and blender of chemicals that contains information about chemical composition, physical and chemical properties, health and safety hazards, emergency response, and waste disposal of the material.

Musculoskeletal Disorders (MSD) — Injuries or illnesses that affect the muscles and skeleton of the body. Defined as Class B medical conditions and may include arthritis, osteoarthritis, fractures, or dislocations among others.

N

N-95 Filter — Certified HEPA filter for use in medical emergencies.

National Association of Emergency Vehicle Technicians (NAEVT) — Professional organization for personnel who maintain emergency vehicles and apparatus. Provides continuing education and certification for maintenance personnel.

National Fire Protection Association (NFPA) — A nonprofit educational and technical association devoted to protecting life and property from fire by developing fire protection standards and educating the public.

National Institute of Safety and Health (NIOSH) — U. S. Government agency that helps ensure the safety of the workplace and associated equipment by conducting investigations and making recommendations.

National Safety Council — Nonprofit organization that collects data on safety and develops and provides safety training courses for the general public.

National Safety Council Defensive Driving Course — Model training program in safe and defensive driving techniques.

Nondestructive Testing (NDT) — Method of testing metal objects that does not subject them to stress-related damage.

O

Occupational Safety and Health Administration (OSHA) — U. S. government agency that develops and enforces standards and regulations for safety in the workplace.

Offensive Strategy or Mode — Aggressive, usually interior fire attack that is intended to stop the fire at its current location.

P

Personal Alert Safety System (PASS) — A motion detector worn by firefighters to alert others in the event the firefighter becomes incapacitated.

Personal Protective Clothing — Equipment designed to protect the wearer from heat and/or hazardous materials contacting the skin or eyes. Usually includes helmet with faceshield or goggles, coat, pants, boots, gloves, and hood.

Personal Protective Equipment — Equipment provided to shield or isolate a person from the chemical, physical, and thermal hazards that can be encountered at an emergency incident. Personal protective equipment includes both personal protective clothing and respiratory protection. Adequate personal protective equipment should protect the respiratory system, skin, eyes, face, hands, feet, head, body, and hearing.

Personnel Accountability System — System that readily identifies both the location and function of all members operating at an incident scene.

Pneumatic Tools — Power tools that operate on the principle of air pressure in a closed circuit system.

Portable Equipment — Tools and equipment such as hoses, ground ladders, and smoke ejectors carried on fire apparatus that are not permanently attached to or part of the apparatus.

Positive Pressure Ventilation (PPV) — Forced ventilation technique that uses the principle of creating pressure differentials. High-volume fans force air into the structure, increasing the internal pressure and forcing smoke out of the structure.

Postincident Analysis — Analysis of information regarding the cause of an incident, the department's actions in bringing the incident under control, and the effects of the incident.

Postincident Critique — Objective analysis and review of the events surrounding an emergency incident. Intended to isolate the need for changes in strategy, tactics, or safety policies.

Proximity Clothing — Protective apparel that allows the wearer to get closer to heat and flame than possible when wearing structural protective clothing. Used for fighting very intense fires.

R

Rapid Intervention Crews or Teams — Emergency personnel who are equipped and designated to make an emergency entrance into the hot zone for the purpose of rescuing other emergency personnel. This team or crew has no other assigned tasks while assigned this function.

Records Management and Data Analysis — Function of the health and safety officer that involves the collection and analysis of records relating to all incidents (accidents), occupational deaths, injuries, illnesses, and exposures.

Regulation — Authoritative rule issued by an executive authority of government; procedures, rules, or orders that have the force of law.

Rescue line or Ropes — *See* Life Safety Line or Ropes.

Risk Avoidance — Risk control technique involves avoiding the activity that creates or causes the risk.

Risk Control — Component of risk management that involves the implementation of solutions to reduce the actual or potential hazard. This is the action step in the process.

Risk Evaluation — Component of risk management that determines the likelihood of occurrence of a given hazard and the severity of its consequences.

Risk Identification — Component of risk management that identifies the actual or potential hazards that may be encountered.

Risk Management — Identification and analysis of exposure to hazards, selection of appropriate risk management techniques to handle exposures, implementation of chosen techniques, and monitoring results in respect to the health and safety of members.

Risk Monitoring — Component of risk management that evaluates the effectiveness of the risk control techniques.

Risk Prioritization — Component of risk management that involves placing hazards on a list based on the high probability of occurrence and severity.

Risk Reduction — *See* Risk Control.

Risk Transfer — Risk control technique that physically transfers the risk to someone else or to an insurance carrier.

Royal Canadian Air Force Plan 5BX for Men (1962) — One of the early physical fitness programs adopted by fire departments; it has been replaced by more effective programs.

S

Safety and Health Policy — Written policy that establishes the infrastructure for the fire department's occupational safety and health program. The purpose of the policy is to clarify the member's responsibility in this process as well as the department's responsibility.

Selection, Care, and Maintenance (SCAM) Document — NFPA publication intended to expand on the individual standards and provide more information for implementation of the standard. Similar to the information found in the appendix of most standards.

Service Testing — Regular, periodic inspection and testing of apparatus and equipment according to an established schedule and procedure to ensure that it is in safe and functional operating condition.

Sick Building Syndrome — Condition that occurs within a structure that results in more than 20% of the occupants suffering from adverse health effects.

Southern Area Fire Equipment Research (SAFER) — Voluntary, nonprofit organization consisting of industry and fire service personnel. The organization's goal is to provide improvements in the protective clothing and equipment used by the fire service.

Standard — Criterion documents that are developed to serve as models or examples of desired performance or behaviors. No one is required to meet the requirements set forth in standards unless those standards are legally adopted by the authority having jurisdiction in which case they become law.

Standards Council of Canada (SCC) — Verification agency for standards writing agencies in Canada. Similar to ANSI in the United States.

Station Wear — Fire service work uniforms that meet the requirements of NFPA 1975 and provide protection in the event of an unexpected exposure to fire. Garments are also designed so that they do not contribute to a fire injury on the wearer.

Statutory Laws — Civil or criminal law that is enacted by the legislative body of the authority having jurisdiction.

System Safety Program — Risk management program developed by the National Aeronautics and Space Administration for hazard identification and resolution.

T

Task Analysis — Analyses of tasks or jobs that are performed by fire department personnel. Used to develop physical fitness programs, establish hiring criteria, and design facilities and apparatus.

Three-Sixty — Slang term for a 360-degree inspection of an incident scene made by the incident safety officer.

Triage — System used for sorting patients to determine the order in which they receive medical attention.

U

United Laboratories (UL) Approved — Independent research and testing laboratory that certifies equipment. Equipment is approved only for the specific use for which it is tested.

Utility Ropes — Ropes used in any situation, except for life safety applications, that requires a rope. Utility ropes can be used for hoisting equipment, securing unstable objects, and cordoning off an area.

W

Wind Chill Factor — Index of chart used to determine the effects of cold air temperature when combined with the speed of the wind. Used as a guide for determining the potential hazard of working in cold, windy weather.

extension cord hazards, 96, 113
exterior area of facilities hazards, 99
extinguishers, 126, 127
extinguishing agents available during training, 68
eye protection, 150–151
eyewash stands, 85

F
fabrics
  used for station wear/work uniforms, 145–146
  used in apparatus, 172
facepieces
  cleaning and disinfecting, 85, 89
  testing fit of, 138
facial hair and proper fit of facepieces, 138
facilities safety
  apparatus bay hazards, 97
  building material hazards (asbestos), 99
  cleaning and disinfecting areas, 85–86
  comparing department statistics, 78, 79
  design issues, 78–84
  electrical safety, 96
  ergonomics, 100–101, 161–162
  exterior area hazards, 99
  health hazards, 85–93
  housekeeping and, 94, 98, 113
  indoor air pollution, 86–88
  infection control, 77–78, 92–93
  inspections of stations, 102, 223–224
  job performance requirements for, 76
  kitchen areas, 92, 96, 98
  lighting (illumination), 94–95
  mechanical space hazards, 99
  natural disasters, 101
  noise pollution, 95–96
  office area hazards, 97
  personal hygiene, 89–91, 92
  physical hazards, 93–99
  reducing illnesses and injuries sustained at stations, 56–57
  remodeling and renovating existing facilities, 83–84, 87, 89–90, 95–96
  shop and maintenance areas, 97, 111
  slipping, tripping, falling, 85, 93–94
  statistics on, 78
  visitors and, 101–102
  water quality, 88–89
falling hazards, 85, 93–94
fatalities (deaths), 3, 4, 78, 157–158. See also firefighter safety
Federal Aviation Administration (FAA), 23–24, 170
Federal Motor Vehicle Safety Standard ,121 174
FIERO (Fire Industry Equipment Research Organization), 141
finance units and personnel in Incident Management System command structure, 185
fire behavior considerations at scenes, 191–193
fire code officials and facilities design, 80
fire control tactics, spotting hazards during, 186–187, 197–198
firefighter safety. See also topics generally in this index

deaths (fatalities), 3, 4, 78, 157–158
  ensuring health at incidents, 32
  riding on apparatus, 170
  statistics on deaths and injuries, 1, 3, 4, 78, 157–158
Fire Industry Equipment Research Organization (FIERO), 141
fire load hazards, 187
Fire Service Joint Labor Management Wellness/Fitness Initiative, 37. See also health maintenance
fire suppression, monitoring, 33
fire tetrahedron, 191–192
first aid kits, 56, 57
Flammable and Combustible Liquids Code (NFPA 30), 93
flammable materials
  dispensing, 124
  policies for use of, 123–124
  storing, 93, 125–126
  using during training, 68
  welding operations near, 128
flammable ranges of fuels, 192
flashover stage of fire development, 192
flexibility component of physical fitness programs, 45
foot protection, 109, 147
forcible entry
  hazards during, 187
  in preemployment tests, 41
  tools used for, 116–117
frequency of risks, 21, 215–216
fuels, forms and classifications of, 192
fuel trucks, 171, 173

G
gasoline-powered equipment, 120–121
generators for auxiliary power, facilities safety and, 96
groups at scenes defined in the Incident Management System (IMS), 186
growth stage of fire development, 192
Guide to Developing and Managing an Emergency Service Infection Control Program, 172

H
hand tool hazards, 111–112
harnesses, life-safety, 148–149
hazardous materials
  asbestos, 99
  cleanup and disposal of, 22
  codes and regulations applying to, 4, 51–52
  dispensing, 124
  exposure protocols developed by safety officers, 52
  personal protective clothing for, 142–144
  policies for use of, 123–124
  protecting personnel from, 33
  requirements for handling, 4
  safety officers roles in operations involving, 201–202
  storing, 125–126
hazards to health at facilities, 85–93
health and safety committees. See department health and safety committees

health and safety officers. See department health and safety officers
health and safety programs. See also occupational safety programs
  components of, 1-2. See also names of specific components
  developing, 1
  as implementation of control measures, 23
  reasons for establishing, 9, 25
  safety officers responsibilities defined by, 9
health and wellness programs
  ergonomics addressed in, 100–101
  physical fitness component, 44–48
health maintenance. See also department physicians; medical evaluations; physical fitness
  ensuring compliance to requirements for, 30
  ergonomics, 100–101, 161–162
  musculoskeletal disorders and, 100
  personnel records on, 41, 44
  statistics on physical fitness and wellness, 37
health of personnel at incidents. See firefighter safety
hearing protection, 107–109, 113, 151–152. See also noise pollution
heat stress, 140, 144, 189
helmets, 150, 151, 195
high efficiency particulate air (HEPA) filter-equipped masks, 136
high-hazard occupancies in communities, identified during pre-incident planning, 193
highway operations, personal protective equipment for, 154
hose drags included in preemployment tests, 41
hot zones at scenes, 190, 191
housekeeping and physical hazards, 94, 98, 113
hydraulically powered tools, 114–115
hygiene, facilities safety and, 89–91, 92

I
ignition stage of fire development, 192
illness. See also firefighter safety
  facilities safety and, 78
  medical evaluations performed after lost time, 42
  taking sick leave when necessary, 89
  treating at incident scenes, 55–56
illumination and physical hazards, 94–95
incidence rate of risks, defined, 21
Incident Commanders (ICs)
  ensuring operations match incident action plan strategies, 198
  in Incident Management System command structure, 185
  responsibilities at scenes, 183, 194
  risk management performed by, 199
  safety officers
    assisting, 33
    reporting to, 32, 53, 194–195, 196, 198
    reporting to during training, 67
  strategies for fire fighting decided by, 197

Indexed by Kari Kells

COMMENT SHEET

DATE _____ NAME _____

ADDRESS _____

ORGANIZATION REPRESENTED _____

CHAPTER TITLE _____ NUMBER _____

SECTION/PARAGRAPH/FIGURE _____ PAGE _____

1. Proposal (include proposed wording or identification of wording to be deleted),
      OR PROPOSED FIGURE:

2. Statement of Problem and Substantiation for Proposal:

RETURN TO: IFSTA Editor   SIGNATURE _____
                  Fire Protection Publications
                  Oklahoma State University
                  930 N. Willis
                  Stillwater, OK 74078-8045

Use this sheet to make any suggestions, recommendations, or comments. We need your input to make the manuals as up to date as possible. Your help is appreciated. Use additional pages if necessary.

Your Training Connection.....

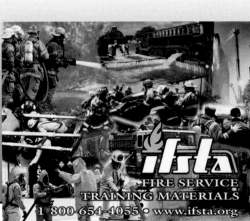

We have a free catalog describing hundreds of fire and emergency service training materials available from a convenient single source: the International Fire Service Training Association (IFSTA).

Choose from products including IFSTA manuals, IFSTA study guides, IFSTA curriculum packages, Fire Protection Publications manuals, books from other publishers, software, videos, and NFPA standards.

Contact us by phone, fax, U.S. mail, e-mail, internet web page, or personal visit.

Phone
1-800-654-4055

Fax
405-744-8204

U.S. mail
IFSTA, Fire Protection Publications
Oklahoma State University
930 North Willis
Stillwater, OK 74078-8045

E-mail
editors@ifstafpp.okstate.edu

Internet web page
www.ifsta.org

Personal visit
Call if you need directions!